W9-BDH-727

A Field Guide to

HOUSEHOLD
TECHNOLOGY

ED SOBEY

CHICAGO
REVIEW
PRESS

Library of Congress Cataloging-in-Publication Data

Sobey, Edwin J. C., 1948–

A field guide to household technology / Ed Sobey.

p. cm.

Includes index.

ISBN-13: 978-1-55652-670-1

ISBN-10: 1-55652-670-9

1. Inventions—History—20th century. 2. Household appliances—Technological innovations. 3. Household electronics—Technological innovations. I. Title.

T20.S495 2007

683'.8—dc22

2006029494

To Molly

Cover and interior design: Joan Sommers Design

Published by Chicago Review Press, Incorporated
814 North Franklin Street
Chicago, Illinois 60610
ISBN-13: 978-1-55652-670-1
ISBN-10: 1-55652-670-9
Printed in the United States of America
5 4 3 2 1

CONTENTS

ACKNOWLEDGMENTS

Steve McCracken, an Eastside Running friend, provided information on the telephone system inside the house. Ryan Collay, a friend at Oregon State University, suggested several of the topics that are included in this book. Rod and Joyce Brown graciously allowed me to track mud through their home in a quest for photos. Rod also took the photo of his office water cooler. Neighbors Michael and Katrina showed me the awesome technology they've installed in their home. Both being engineers, they have the bells and whistles sounding sweetly. Spike Cowen showed us his central vacuum system and allowed us to wander around his home taking photos. Another running friend, Carl Kadie, offered his home for photos. Perry Rodgers, yet another running friend and pinball wizard, shot the photo of his cable box for me.

John Dickson, professional photographer and runner, snapped the iPod photo, and Phil Block took the shot of TiVo, a device he loves. My son, Andrew, snapped several photos including the trash chute at a dorm at Central Washington University.

Forrest M. Mims III, a highly regarded scientist and prolific writer, graciously provided the photo of the lightning rod. Valerie Reddemann of Greenfeet (www.greenfeet.com) provided the laundry chute photo. Connecticut Screen Works, Inc. allowed me to use the photo of one of its shades. Ken Wright, with Hot Sun Industries Inc. (www.powermat.com)

provided the photo of a solar hot water heater for a swimming pool. You might imagine how difficult it is to find a pool solar heater in the state of Washington—thanks, Ken.

Barbara, my wife, put up again with the craziness of writing yet another book and provided her good suggestions on what to cover. I thank her and all the others who entertained unusual requests for information and photos.

"What is that thingy on the wall?"

"How does a dimmer switch work?"

"Why are there funny outlets in the bathroom?"

Technology isn't all connected to your computer. It's in every widget and gizmo that you use.

Every device inside and part of your home represents the collective efforts of hundreds or thousands of engineers, designers, and scientists who have figured out how to provide convenient, safe, and affordable services and solutions to you. In many cases, the same designs show up in many very different appliances. The bi-metallic thermostat shows up on your wall to regulate house temperatures as well as in your toaster. When a machine or component part has evolved to its most effective level, manufacturers use it in many different devices. Once you recognize the component in one gizmo, you'll recognize it in others as well.

Although we rely on many devices inside our homes, we are aware of only a few, and are aware of them mainly when they fail. *A Field Guide to Household Technology* shows you what this stuff is and gives concise descriptions of how it works. With this book in hand, you can investigate every nook, cranny, and dusty recess of your home to discover the wondrous machines that play such important roles in your life, yet are hidden from view and consciousness.

Discovering these gizmos and how they work opens the world outside your home to your questioning. How does electricity get from a hydroelectric generating plant miles away and where do the sewage flushings go? Learning is such a wondrous and never-ending venture, I hope this book helps propel you to learn more.

Like its companion book, *A Field Guide to Roadside Technology*, this book is organized in habitats. Each room or type of room includes its own suite of technologies, and you can find the device of your inquiry in the chapter corresponding to the room where the device is usually found. Look at chapter 4 on the kitchen to find information on dishwashers, microwave ovens, and can openers.

Some systems span the geography of the house and are listed separately. These systems are listed in their own chapters: "Heating and Air Conditioning," and "Lighting and Electrical Systems." For example, lightbulbs are found in every environment of a house, so I list them in the "Lighting and Electrical Systems" chapter. However, where a component shows up only at one location, I list it at its location rather than in these two systems chapters. For example, electric meters could be in the "Lighting and Electrical Systems" chapter, but I chose instead to list it where you would find it—so it's in the chapter on "Patio, Porch, Roof, and Outside." GFI switches occur predominately in bathrooms, so they are listed there.

Each entry includes a description of the behavior of the device (what it does); its habitat (where you find it); and how it works (the technology). Many also include things to look for (distinctive features you can see) and other neat things (information on the development of the technology or related technology).

Although admittedly weak in plot and character development, this book might be compelling enough for you to read cover to cover. You might enjoy leafing through the book and developing a greater appreciation for the inherent engineering beauty embedded in the common devices. Or, you might enjoy jumping from room to room seeing what stuff is listed where.

My intention, however, was to create a guide book to answer the questions "What's that?" and "How does that work?" This is a reference book for the technology you bump into at home. Much like a field guide that identifies and describes the trees and rocks you might bump into outdoors, *A Field Guide to Household Technology* is meant to be always accessible, ready to answer questions.

A secondary, but possibly more important role for *A Field Guide to Household Technology* is to help you ask questions. I want to encourage you to look, ask, and think about the stuff that surrounds you. You might indeed make an important discovery enabling you to invent the electric potato peeler or four-way lightbulb. Or maybe you will discover a curiosity that requires much more research.

This field guide enables you to satiate your curiosity quickly by leafing through the pages in the appropriate habitat chapter and finding either the photograph or description. However you use the book, I hope you use it often.

ENTERING THE HOME

HOW DO YOU KEEP unwanted people out of your house while letting the family in? This problem is as old as buildings; once people put up walls to keep out intruders, they had to have doors to control egress, and systems to secure the doors.

In modern homes we use a variety of devices. Some, like locks, have roots in the ancient world and others are newcomers. With the miniaturization of electronics and development of new sensors, entering the home has gotten a lot more technical than it was even a few years ago.

Door Chimes

BEHAVIOR
Alerts you to visits from Aunt Kay, the UPS driver, and Girl Scouts selling cookies.

HABITAT
The button that activates the chime is mounted on the outside doorway of the front door. The chimes themselves reside indoors, usually high on a wall adjacent to the doorway.

HOW IT WORKS
Doorbells and chimes rely on electromagnets to operate. When Aunt Kay pushes the doorbell button, she completes an electric circuit that sends current through a transformer (to change the voltage) to an electromagnet inside the bell or chime box. The electromagnet becomes magnetized when current flows through it, and the now-energized magnet pulls a metal arm that strikes the bell or chime. When the current is interrupted, the electromagnet loses its power and a spring pulls the arm back.

In chimes, the electromagnet is part of a solenoid, or simple electric motor. A solenoid moves back and forth, unlike most motors, which move circularly. With the button depressed, the solenoid moves forward striking a chime (metal bar) that plays a note. When the button is released, the solenoid is pulled back by a spring so it hits a second chime on the opposite side. Together the two chimes give the "ding dong" sound.

On buzzers and bells, the electromagnet propels an arm to strike the metal bell. As the arm moves, it interrupts the electric circuit. With the circuit now open, the only force on the arm is a spring that pulls it back to its starting position. Now back at its initial position, the arm makes electrical contact again, re-energizing the electromagnet, which pulls the arm to strike the bell, over and over again.

Storm Door Closer

BEHAVIOR
It pulls the storm door closed, but allows it to close gradually.

HABITAT
Attached to the inside of storm doors. One end is screwed into the door and the other is screwed into the door frame.

HOW IT WORKS
Before this type of closer was used, long springs pulled screen doors closed. This tended to slam the door shut, which often caused it to rebound open and slam shut again. And again.

To prevent this "slam-slam" annoyance, and to prevent damage to the glass in a storm door, pneumatic closers are often installed. Inside the metal tube is the spring that pulls the door shut. A piston is inside. As the door is opened, the piston extends and draws air into the tube through an opening. Releasing the door allows the spring to pull it shut, but the piston is now pushing air out of the tube through a small hole that resists this motion. The air opening has an adjustable valve so you can control how quickly the door closes.

The closer also has a stopper so you can prop the door open. This is very handy when you're hauling in bags of groceries. The metal stopper slides on the piston shaft. Setting it at an angle wedges it into the shaft to hold the door.

A safety chain and spring prevents the door from opening so far that the closer or the door hinges get damaged.

Front Door Lock

BEHAVIOR
Keeps out (some of) the people you don't want inside, but gives those you do want inside easy access.

HABITAT
Mounted on the side of the door opposite to the hinged side. Located at a convenient height to allow users to insert a key to open the door.

HOW IT WORKS
Most doors use a cylinder lock. Inserting the correct key allows you to rotate the cylinder, which is connected to an arm that withdraws the latch keeping the door secure. The latch is usually pushed closed with an internal spring.

The beauty of the lock is that each has its own code that protects it. The code is cut into a metal key. The vertical indentations in the key correspond to both the placement and heights of pins inside the lock. As the key enters, it pushes the spring-mounted pins up and out of the way. A key with the incorrect code will not push the pins to the height that allows the cylinder to turn.

From inside the house you can operate most locks by turning a knob. This is a nice safety feature so you don't have to find a key to escape a fire. On the other hand, for higher security from intruders, some locks require keys on both the inside and outside to open. The inside key prevents someone from breaking an adjacent window and reaching through it to open the door. It can also prevent people— toddlers or burglars who entered another way—from exiting the door.

> The first lock system that used keys was created in Sparta around 400 B.C.

Security Door Viewer or Peephole

BEHAVIOR
Gives you a wide-angle view of your front porch area from behind the front door.

HABITAT
Found at eye level, or a bit lower to accommodate shorter people, in the middle of the front door.

HOW IT WORKS
The optics in the peephole work like a wide-angle lens. The glass refracts or bends light to collect images from a wide swath and focus them into the small viewing port. Some peepholes have large viewers so you can see outside from several feet back from the door. Other peepholes even allow you to connect a video camera.

Standing outside and peering in, you can't see much at all. Of course, some creative people make a viewer (for law enforcement agencies) that fits on top of the outside of a peephole and allows them to see inside.

Keypad for Home Security System

BEHAVIOR
Allows the owner to enter a code to arm and disarm the security system.

HABITAT
The keypad is located inside the house, near the entry door used most often. It is inside to protect it from tampering. Being inside requires that it have a programmed delay (about 30 to 45 seconds) so the owner can enter the house, turn on a light, and disarm the system by punching in a code before it triggers an alarm.

HOW IT WORKS
A microprocessor inside the keypad allows you to set and change the code. It also allows you to turn on parts of the system rather than ener-gizing the entire system. For example, if you were alone inside the house, you could activate the perimeter alarm (to alert you that some-one had entered the house), but not the motion detectors. This would provide security without your movements inside the house setting off the alarm. Or you could activate the alarms in one part of the house but not the entire house.

Although the typical five-digit code used in many alarms would be easy to crack, intruders only have a few seconds to enter, find the key-pad, and try a few codes before the alarm sounds.

The keypad can activate local alarms or send alarms to off-site secu-rity companies or even police departments. Some systems integrate entry alarms with fire alarms.

Perimeter Entry Detector or Burglar Alarm

BEHAVIOR

It alerts you or a security company that someone has entered the house through a door or window.

HABITAT

On homes with this type of security system, look for detectors with wire leads on doorsills and alongside windows.

HOW IT WORKS

There are several types of perimeter alarm detector, but most common is one that uses magnetic reed switches. A magnet is attached to a door or window. When the door or window is closed, this magnet lies adjacent to a magnetic switch attached to the doorframe or window frame. With the alarm armed, if someone opens the door or window, he moves the magnet, which was holding the switch in the closed position. Now free of the magnet's pull, the switch opens and causes the alarm to sound.

Perimeter alarms make it difficult for an unwanted visitor to gain access to your home. If a visitor does gain entry, security systems have a second level of alarms that detect motion within a room.

INTERESTING FACTS

Stores use perimeter entry detectors to let the clerks know that someone has entered the store. These typically use visible or infrared beams of light that cross the entry of the store, just below knee height. When you enter, you interrupt the beam of light and that causes a "gong" or bell sound. Usually these devices are clearly visible, mounted on each side of the door.

Perimeter Alarm Switch for Doors

BEHAVIOR
Sends a signal whenever the door has opened.

HABITAT
Built into the door frame, where it is very difficult for burglars to find and disconnect it.

HOW IT WORKS
This is a simple spring-loaded on/off button that is installed into the doorframe. When closed, the door holds the button in its compressed position. When the door opens, the internal spring pushes the button out signaling that the door is open. With the alarm turned off or with that zone of the alarm system turned off, opening the door doesn't trigger an alarm.

INTERESTING FACTS
This simple electrical system is unlikely to fail. Since intruders can't see the alarm switch from outside (or from inside for that matter, unless the door is open), they can't disarm it.

Motion Detector

BEHAVIOR

It senses your cat wandering around the house at night or, possibly, that burglar you've always feared.

HABITAT

You can spot these devices hanging high on a wall, often in the corner of a room. They are usually the same (approximate) color as the wall so they blend in. They have what appears to be a curved lens, and some have an active LED (light emitting diode) to indicate that the sensor is working.

HOW IT WORKS

Most motion detectors use infrared imaging. All objects emit infrared radiation; the sensor detects major changes in the amount of infrared radiation in the room. It sends an alarm when it detects a significant change in the infrared energy. Someone entering an empty room will add infrared radiation that the detector can distinguish.

The combination of perimeter and motion detectors makes it quite difficult for anyone to enter undetected. Unless someone wants to go to the lengths of *Mission: Impossible* to get in, it's easier to find a less well-protected home.

Video Monitoring Camera

BEHAVIOR
Allows occupants to see who is at the front door from any television set in the house.

HABITAT
Found at gated developments and at front porches. It is positioned to give the best view of anyone standing at the door.

HOW IT WORKS
The video camera sends its signal to a modulator that is probably mounted on a wall in the utility room, basement, or closet. The modulator converts the signal to a television UHF (ultra-high frequency) channel that the user selects.

When someone rings the doorbell, the user changes the television channel to the selected UHF channel to view the video coming from the front porch. If the user has an intercom system, he or she can talk to the magazine seller waiting to give the pitch, and can send the seller away without coming to the front door.

Garage Door Opener

BEHAVIOR
Lets you open the garage door without getting out of your warm, dry car and walking through the cold, drenching rain.

HABITAT
This remote control is found on the visors, dashboards, and glove compartments of many cars. The receiver and motor are mounted from the ceiling of the garage.

HOW IT WORKS
Unlike the TV remote control, the garage door opener uses radio waves, not light. Older systems use tiny switches to encode a signal into the transmitter (in the car) and receiver (adjacent to the motor that opens the door). By encoding the signal, your neighbor (hopefully) will not open your garage door when trying to open his.

Newer systems have much more elaborate electronic systems for encoding the signals to prevent burglars from capturing the signal and entering your house. Transmitters and receivers generate new codes each time they function, thus making it very difficult for someone to swipe a code.

For more information on the garage door opener itself, see page 142.

Garage Door Opener Touch Pad

BEHAVIOR
Allows you to open your garage door without the handheld remote control unit kept in your car. Coming back from a bike ride or otherwise locked out of your house without a front door key, you can enter your personal code to open the garage door.

HABITAT
Mounted on the doorframe of the garage door or on an outside wall.

HOW IT WORKS
Punching in the code (which you can usually set yourself) that is recognized by the logic circuit in the garage door electronics opens the door. The opener is wired to the electronics, which are housed adjacent to the motor near the ceiling and near a power outlet.

Keyless Entry Device

BEHAVIOR

Allows you to lock and unlock your car doors and trunk without a key. Also allows you to find your car in a crowded parking lot when you've forgotten where you've parked it. This is an especially valuable function when renting a car.

HABITAT

Found amid pocket lint in the pockets of many Americans. Usually found on a key chain or ring with a set of traditional keys, just in case the battery inside the remote entry device should die.

HOW IT WORKS

A computer inside your car monitors the several different kinds of signals that you might send to unlock the doors. You could enter a code on the outside of the driver's door, push a button on the inside of either front door, or push a button on your keyless entry device or fob.

When the car computer detects the radio signal from your fob and determines that it is the right signal (and not the guy parked next to you trying to get into his Lotus), it powers a motor or actuator. The actuator, which has a small DC motor and gears inside, only moves up or down. Moved up, the actuator pushes a lever that connects the outside door handle to the lock, allowing you to open the door. When you push "Lock," the actuator moves down and disengages the outside door handle from the lock. No matter how hard you pull on the handle, it won't open the door.

Intercom

BEHAVIOR
Allows you to play a radio station on speakers throughout the house and to talk to other people throughout the house or at the front door.

HABITAT
Usually there is a speaker outside the front door and in several rooms. A control box with a radio built-in is usually located in the kitchen or a central hallway or entryway.

HOW IT WORKS
Older home intercoms had units throughout the house wired to a master unit. Newer systems are wireless.

Wireless units (not requiring installation of wires) either operate as broadcast radios or send signals through your existing electric wires.

Some intercoms come with video screens, so you can see the person standing at the front door selling magazine subscriptions.

LIVING ROOM, FAMILY ROOM, AND DEN

ENTERTAINMENT SYSTEMS DOMINATE the technology of these living spaces, and entertainment systems typically run on electrical power. So, much of the technology found here is found plugged into electric wall outlets.

Of course, not everything in these rooms is related to entertainment. No one would consider a vacuum cleaner a source of amusement.

Vacuum Cleaner

BEHAVIOR
Sits quietly until company is about to arrive. Then it roars to life, lifting and sucking dust bunnies and dirt from carpets and hardwood floors.

HABITAT
When not in use, stored in a hall closet. In the hands of a frantic cleaner, it can be used almost anywhere in the house.

HOW IT WORKS
Vacuum cleaners are marvels of engineering. A motor drives a fan that pulls dirt-laden air inside. It pushes the air into a porous collecting bag that traps the dirt. When full, the bag is removed, thrown away, and replaced with a new one.

In bagless vacuum cleaners the dirt-laden air is blown into a cylinder. The heavy dirt moves to the outside of the cylinder as the air flows in a vortex, and the clean air escapes upward and out the center of the cylinder. The collected dirt is emptied through a trap at the bottom.

The vacuum's motor also operates rotating brushes underneath that flick dirt up so it can be caught in the airstream and transported to the bag. A rubber belt connects the motor to the brushes and, with age, this belt can break.

INTERESTING FACTS
Before vacuum cleaners were invented, there was the blower cleaner. It blew dirt into a collecting box. And it didn't work well, but inspired Cecil Booth to invent the vacuum cleaner. He first used his mouth to suck dirt from the back of a chair in a restaurant and was so successful that he almost choked. His first mechanical vacuum was so large that he mounted it on a horse-drawn cart. Rather than sell the vacuum, he used it to provide a cleaning service.

Central Vacuum Cleaner

BEHAVIOR
Like an iceberg, 90 percent of this system is hidden from view. All you can see, unless the vacuum is in operation, are the inlets mounted on the walls of every room.

HABITAT
Found in about 4 percent of U.S. homes, the vacuum system resides in the basement or garage. The pipes are built within the walls with inlets mounted on walls about a foot off the floor.

HOW IT WORKS
The central unit provides suction and filtering (see chapter 7). With inlets closed except the one attached to the working cleaner attachments, the central unit sucks up dog hair, crumbs of chocolate cake, and dust bunnies and pulls them into a collecting bag that you empty periodically. A variation is the bagless container that spins the dust-laden air, dropping heavy particles into a bin (that you empty), and discharges the air outside. Filters allow air molecules to pass through, but not the much larger particles of dirt. Periodically the filters must be cleaned or replaced.

The advantage of a central vacuum system is not having to lug the vacuum cleaner itself around and not having to plug it into an electrical outlet and get tangled in the cord. Some systems allow you to install dustpans at floor level (see chapter 4). Open the door and use a broom to sweep dirt directly into the system to be sucked away.

INTERESTING FACTS
Because central vacuum units are much larger than the ones you pull or push, they hold a lot more dirt and require emptying only a few times a year. These systems can be retrofitted into existing homes as well as installed during construction.

Robotic Vacuum Cleaner

BEHAVIOR
Acts like an intelligent device, moving around a room, sucking up dirt and debris.

HABITAT
Found in the homes of technology fans.

HOW IT WORKS
Not so smart that it can find dirt and go after it, robotic vacuum cleaners wander around in programmed patterns. Two motors drive the robot, one for the left drive wheel and the other for the right. Some models bump into walls and objects to detect them and then head off in a new direction. Newer models use infrared sensors to detect and avoid objects. Another sensor detects steep drop-offs to stop the vacuum from plunging down stairs.

Side brushes push dirt in toward the robot and an agitator brushes dirt beneath it into the dirt bin. A vacuum also sucks dirt into the bin. The bin is small and needs to be cleaned after each room is vacuumed. The onboard battery allows a couple of hours of operation before recharging.

Commercial success of early models ensures that, as new technology develops, smarter vacuums will emerge.

Carpet Sweeper

BEHAVIOR
Picks up dust and dirt without having to be plugged in to a wall outlet.

HABITAT
Lives in the hall closet or utility room and sees the light of day typically only after a small accident has occurred and cookie crumbs or dust bunnies need to be corralled.

HOW IT WORKS
Imagine an appliance that doesn't require electricity! You supply the power to make the carpet sweeper work. As you push the sweeper back and forth across the carpet, the wheels turn drive belts that rotate brushes. The brushes push dirt into the holding box that needs emptying periodically.

INTERESTING FACTS
Often called a Bissell regardless of their brand, these were invented in the early 19th century. Melville Bissell, allergic to straw dust, improved earlier designs of carpet sweepers to reduce the dust in his china shop. His machine worked so well that he and his wife formed the Bissell Carpet Sweeper Company. The "Bissell" became a popular American home appliance.

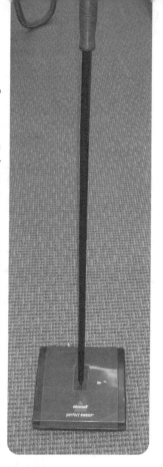

Compact Disc (CD)

BEHAVIOR
When inserted into a computer or CD player it makes accessible an unbelievably dense warehouse of information, music, or photographs.

HABITAT
Prized CDs are lovingly stored in plastic wrappers kept in three-ring notebooks or in plastic boxes. Of course, the most favored are kept in the CD player, ready for instant access. The least favored (commercial promotions) are taken immediately from the mailbox to the trash bin.

HOW IT WORKS
The CD player or computer does all the heavy lifting. But the CD itself is an ingenious device onto which millions of divots are laser-etched, each one part of a digital signal.

The CD is 120 mm in diameter and 1.2 mm thick. It holds either 74 minutes of audio recordings or 650 megabytes of data. The disk is made of polycarbonate plastic covered with a fine layer of aluminum and protected with a coating of lacquer. Information is etched into the aluminum (on the underside of the CD, not the side with the label) using a laser and is read by bouncing a lower-power laser beam off the surface. Unlike the vinyl records it replaced, a CD records data on spiral starting nearest the center opening and records outward. The track is about 3.1 miles (5 km) long. The etches made in the CD by the recording laser are 100 nanometers deep (a nanometer is one billionth of a meter). To clearly see a pit, you'd have to enlarge a CD so its diameter is greater than a kilometer.

The reading laser bounces light off the surface of the CD. Bounced off a pit, the reflected light is greatly reduced in strength. The change from a pit to "land" (no pits) and from land to pits denotes a binary one, while no change in the reflected signal denotes a binary zero.

Billy Joel's *52nd Street* album was the first CD recording to be sold.

Compact Disc Player

Spins CDs and reads the music encoded on them so you can hear it.

They're everywhere. Home entertainment centers or stereo systems have CD players. Individual units— Discman or Walkman and other brand names—are carried everywhere.

It takes a laser to read a CD. The tiny diode laser inside the CD player focuses on the spiral track of the CD. The CD spins at a variable rate, depending on where the laser is reading. Reading at the start (center) of a CD, the CD spins at 500 revolutions per minute. As the song progresses, the laser is carried farther toward the outer edge of the CD and the CD spins at a slower rate. If you look inside a CD player you can see the mechanism that carries the laser (look for a lens) in and out.

The laser sends an infrared beam of light toward the CD. The reflected light is detected by photo diodes mounted adjacent to the laser. The depth of the pits or bumps in the CD is engineered to give maximum discrimination between a pit and land (area with no pits), so the photo diodes can read the encoding. The signal of digital bits is fed through the electronics, amplified, and sent to a headphone or speaker.

Digital Video Disc (DVD) Player

BEHAVIOR
Behaves and looks like a CD player. Individual DVDs, although the same size as a CD, play much longer. DVDs use a different format that supports a higher data density.

HABITAT
Found adjacent to television sets, sometimes fixed in entertainment cabinets, and sometimes placed on top of the TV or on the floor.

HOW IT WORKS
The digital video disc (DVD) works like a CD, but it can store about seven times as much data. A DVD has enough data capacity to store a full-length movie.

Both DVDs and CDs are read by lasers. Both are 120 mm in diameter and made of polycarbonate plastic with a thin metal coating. However, DVDs spin about three times faster and use a different lens that can discriminate more densely packed data.

A DVD player uses a laser operating at 650 nm (nanometers, or one billionth of a meter), while a CD player operates at 780 nm. The shorter wavelength for DVDs allows the laser to read finer detail. Data covers more of the surface area of a DVD than CD, and less data storage is allocated to error correction. The various changes from the CD format allow the DVD to hold so much more data.

Like CDs, DVDs are read from the inside (near the center) towards the outer edge in a long spiral. The spiral for a DVD is over 7 miles long (11.3 km).

Television

BEHAVIOR
It projects video signals on a glass screen and plays corresponding audio signals through speakers. The combination captivates humans and causes them to sit through hours of commercials.

HABITAT
Living rooms, dining areas, dens, and even bedrooms are homes to television sets.

HOW IT WORKS
The video signal arrives as three separate color components (red, green, and blue), and each generates a beam of electrons aimed at the screen. The three beams are directed across the screen from top to bottom and from left to right along 525 paths or lines. The beams, however, don't travel along each line in order. They go to every other line from top to bottom and then start at the top again to scan the other lines. This interlacing of the signals gives a new image every 1/60th of a second. To see all the lines of an entire image takes twice as long, 1/30th of a second. By interlacing the signal, your eye sees twice as many images (although only half of each image) in each second, and that gives a smoother flow of images.

The electron beam for the color red hits only the red colored phosphorus dots in the glass screen, which causes a tiny red glow. Beams for the green and blue signals also only hit the dots for their colors. To keep the beams from hitting the wrong color dots, there is a metal screen with tiny holes next to the layer of dots. The screen is called the "shadow mask." The right combination of light from the three different colored dots can make an image of any color.

UNIQUE CHARACTERISTICS
Use a hand lens to examine the TV screen. You'll see the individual color dots that make up the image.

INTERESTING FACTS
Analog (traditional) television broadcasting is mandated by Congress to end by February 18, 2009. By that time, television stations will be broadcasting in digital formats only. Old televisions, without digital adapters, will no longer work with broadcast signals.

Plasma Television

BEHAVIOR

Provides drop-dead color and stunning detail in huge images.

HABITAT

Found in living rooms and dens of nicer homes. If there is an absolute sports nut in the house, there is a good chance of finding a plasma television there.

HOW IT WORKS

A traditional TV shoots electrons from a gun in the back of the tube to the surface where you see the image. This requires the tube to be quite deep. Plasma screens are only a few inches deep and use an entirely different mechanism to create the picture.

The plasma in plasma TVs is electrically charged xenon and neon gas. The gas is held in many thousands of cells. To create light, an electric charge is applied across a cell, which causes the gas inside to ionize and give off light, similar to how fluorescent lights work (see page 186).

The light given off, however, is invisible ultraviolet light. Again, like a fluorescent bulb, the invisible light strikes phosphorescent elements causing them to give off visible light. Each picture element (pixel) is composed of red, green, and blue phosphors (material that gives off light when exposed to other light). By varying the electrical charge at each cell, each phosphor is excited to a higher or lower level, which changes the color.

Because the images are formed in tiny cells, plasma screens are very thin. You can hang them on the wall.

INTERESTING FACTS

Plasma screen TVs use much less electrical power than traditional TVs do.

Put this in your living room: the largest plasma TV measures more than 100 inches (diagonally, across the screen). That's 8 feet across!

Television Remote Control

BEHAVIOR

It allows you to surf the wasteland of TV without arising from the couch.

HABITAT

These devices have the habit of hiding. You can't find one when you need it. Look between the cushions of your sofa or, in a perfect world, on top of the TV set itself.

HOW IT WORKS

Pushing a button on the remote control closes a switch inside that is detected by an integrated circuit. The circuit sends a pulsed signal to a light emitting diode (LED). You can't see the light because it is in the infrared range, not the visible range. However, you can see it if you look at the remote control through a digital camera while operating the remote.

The television set reads the signal and does your bidding. The signals are long binary codes that tell the television that the signal has started, what the command is, which device (TV, VCR, DVD player, etc.) the signal is intended for, and that the signal has finished transmitting.

The infrared signals only work in line of sight. That is, you can't face away from the TV and control it. Unless, of course, the signal reflects off a solid object and bounces back to the TV. TV remotes typically work for only about 30 feet.

UNIQUE CHARACTERISTICS

Try getting the remote control signal to bounce off a wall or book to operate the TV. See how far away you can stand and still change channels.

High-Definition Television (HDTV)

BEHAVIOR
You thought TV was cool? Wait until you see this. The images are much brighter and crisper.

HABITAT
HDTV is found in the prime viewing rooms of a home.

HOW IT WORKS
High-definition television, or HDTV, provides much sharper pictures than traditional TV. It uses 1,050 lines to transmit each picture, twice the number of lines used in traditional television.

The inside of the TV screen is lined with phosphorus dots that light up when hit with a beam of light. There are more than 300,000 phosphorus dots on a HDTV screen. Go ahead, try to count them.

INTERESTING FACTS
Although still rare, HDTV sets are selling well and some predict that they will be very common by 2009.

Antenna Rotator Control Box

BEHAVIOR
Allows television users to rotate the roof-mounted antenna so they can get the best possible signal from the television station.

HABITAT
The control box is found atop the television set and the rotator holds the base of the antenna on the roof, often attached to the side of the chimney. Antenna rotators are typically found in smaller towns and suburbs where television reception is marginal.

HOW IT WORKS
Users turn a dial on the control box to the position optimal for the television station they want to receive. They figure out the optimal direction by trial and error.

The dial connects to an electronic component that switches power on to the motor on the roof. As the rotor reaches the desired position, the circuit is opened, stopping the motor.

Cable Box

BEHAVIOR

It gives your television access to the various channels that are available from your digital cable provider.

HABITAT

It often sits atop the television and plugs into a wall outlet with a coaxial cable (round cable with the outer conductor surrounding the inner conductor).

HOW IT WORKS

The cable company sends its complete set of offerings through optical fiber and then coaxial cable into your home. In analog systems, each channel is allocated 6 Mhz (6,000,000 cycles per second) bandwidth on the cable. Although that sounds like a lot of bandwidth, the cable can carry up to 90 channels and still have room left over for telephone service or surfing the Web. Switching to digital service increases the capacity (potential number of channels) about sixfold.

The signal originates at the "head end" or regional office of the cable company. From there it goes over optical fiber to a "node." At the node, the signal is converted from light waves to electrical signals and sent out on cables, amplified where needed, to the connecter on the utility pole or in the ground pedestal close to your home. A wire taps into the connecter to bring the signal into your home.

Cable TV began as a grassroots effort to get and share television signals in outlying communities. Creative people, frustrated by their inability to receive clear TV signals from distant stations, erected large antennae and figured out how to share the signal with neighbors. Conceptually that's what cable providers do today.

TiVo

BEHAVIOR
Records television programs and plays them back. But, more than just a digital video recorder, it has advanced features that benefit avid TV viewers.

HABITAT
Found perched on top of television set consoles.

HOW IT WORKS
TiVo has a hard drive onto which it records digital video and sound. Users can program TiVo to record a show every time it comes on, or to search the broadcast landscape for a particular actor or type of show to record. Users can record a show and watch it minutes later so they can fast-forward through commercials.

Inside a TiVo is a microprocessor or computer and another component that compresses and uncompresses video signals (so they don't take up as much room to record). The hard drives can range in size from 40 to 250 gigabytes (GB).

Video Game

BEHAVIOR
Engages the eyes and hands of players in a wild assortment of active entertainment.

HABITAT
They can be connected to the home entertainment system or can be small enough to be carried in a pocket.

HOW IT WORKS
If you have a Nintendo, Sega Genesis, or other electronic gaming system, you know they have four components. You watch the action on a screen (possibly your television screen); you control the action with a joystick, hand controller, or other device; the console or game deck is the computer that runs the game; and the game pack, cartridge, or CD has instructions specific to the game you're playing.

The console has many of the same components you would find inside a personal computer. Both have a central processing unit, or CPU, which controls the other components. Both have a component or chip that processes images for display. Both have RAM, or random access memory. In the video game, RAM keeps track of your score. A computer has ROM, read-only memory, built in; in video games the ROM comes in each game CD or cartridge you buy.

INTERESTING FACTS
The first commercially successful video arcade game was Pong in 1972. Nolan Bushnell formed Atari and launched the video game era. He also launched the Chuck E. Cheese restaurant chain and at least 20 other companies.

Radio

It sounds great; or at least it can sound great. It brings you the news, weather, sports, and music.

Radio is one component of the home entertainment system, the personal wake-up (alarm clock) system, and the emergency information (portable radio) system. Radios can be found throughout the house. In some homes, speakers are wired so a radio station can be heard throughout.

Radio stations transfer sound vibrations into electrical vibrations and then convert these into radio waves. They broadcast the electromagnetic radio waves carrying their program from the antennae on tall towers. The waves move at nearly the speed of light (186,282 miles per second, or 299,792 kilometers per second). They travel so fast that if you were listening to a radio broadcast of a baseball game, you could hear the sound of a bat hitting the ball before the player in left field could, even if you and your radio were thousands of miles away. Compared to light, sound pokes along at a lazy 1,000 feet per second

(330 meters per second). That's why you see lightning flash before you hear the thunder's clap.

Although we can't see or feel radio waves as they zip through the air, we can capture them with an antenna, located either inside the radio or mounted outside, such as a car radio antenna. Radio waves striking an antenna cause electrical charges in it to vibrate back and forth. These moving charges are an electrical current that a radio receiver can amplify. The amplified electrical signal is fed into a speaker where it vibrates a piece of paper. The moving paper vibrates air molecules that we hear as sound.

INTERESTING FACTS
Guglielmo Marconi is considered the father of radio as he was the first to transmit a radio signal across the Atlantic Ocean (1901). However, many people contributed to the invention of radio, including Alexander Popov who, in 1896, was the first person to transmit an intelligible radio signal.

For radios, AM stands for amplitude modulation and FM stands for frequency modulation. These terms refer to the method of adding the radio signal to a carrier wave for broadcast. You dial your radio to the carrier frequency, but a steady tone would be a boring program. In AM broadcasts, the strength of the signal added to the carrier wave is changed and these changes (modulations) convey the music or talk.

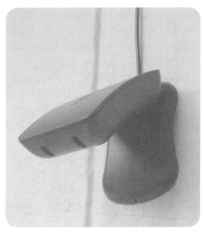

Satellite Radio and Antenna

BEHAVIOR

Brings approximately 100 commercial-free radio stations and a national perspective into your home speakers.

HABITAT

The receiver can be anywhere, but most likely found adjacent to the home stereo or home entertainment system. The special antenna may be mounted on an outside wall.

HOW IT WORKS

The radio signals have traveled a distance equal to twice around the world before arriving at your radio. Two companies, XM Satellite Radio and Sirius Satellite Radio, broadcast their array of sports, news, music, and talk within a frequency band allocated for digital audio radio (2.3 GHz, or 2.3 billion cycles per second).

XM operates two satellites in geostationary orbit. The satellites are approximately 22,000 miles above the earth's equator and move at the same speed as the earth rotates, so they stay in one place (relative to the ground). One satellite is on the longitude of Louisville, Kentucky, and the other is on the longitude that passes through Boise, Idaho. XM transmits its 100 channels of radio up to the satellites and they broadcast them down to radio receivers in cars and homes.

Sirius uses a different broadcast strategy. It uses geosynchronous satellites—they rotate at the same speed as the earth but are not aligned above the equator, so they change position. Sirius has three satellites that move relative to the earth. The movement is designed so at least one satellite is above the United States at all times.

To keep people who haven't paid subscriptions from using the service, each radio receiver has its own serial number. When turned on, it broadcasts this number and receives back from the satellite a code that allows it to receive the radio signals.

UNIQUE CHARACTERISTICS

Look for the distinctive shark fin antenna on the back of the roof of cars. In homes, look for a unit larger than a cell phone that is connected to speakers.

Internet Radio Tuner

BEHAVIOR

Allows listeners to tune in radio stations from anywhere in the world, as long as they are broadcasting on the Web.

HABITAT

Found in the living rooms and offices of Web-centric listeners.

HOW IT WORKS

Radio stations can broadcast on the Web by converting the audio sounds into digital format and sending them to a server that makes them available to the Web. The listener "tunes in" the station (selects the URL of the station) and the tuner or computer converts the encoded signals back into audio and plays them on speakers.

The amazing thing about streaming audio and voice-over-Internet phone calling is that the signal is sent as "packets." Unlike a traditional radio broadcast that is transmitted continuously, Internet radio is broadcast in an interrupted series of bits of data. Each packet can travel from the radio station server to your tuner by a different route. The tuner or computer has to assemble these packets in the right order and play the sounds so you aren't aware that the music isn't coming as a steady stream.

INTERESTING FACTS

Want to catch the Oregon State University Beavers football game while visiting Singapore? Check out the Web site of the radio station that broadcasts the game to see if it is casting it on the Web. More than likely you can catch the game, although it will start at 4 A.M. Singapore time.

Radio-Controlled Clock

BEHAVIOR
Unlike other clocks that have to be reset when the power goes out or when their main spring winds down, this clock resets its time according to a radio station.

HABITAT
You can see these in living rooms, dens, or home offices.

HOW IT WORKS
The National Institute of Standards and Technology in Boulder, Colorado, is responsible for keeping America on time. It broadcasts the current time on a very powerful, very low frequency (60,000 Hz) radio signal that reaches the entire continental United States. It broadcasts the time each minute.

Once a radio-controlled clock is set for its time zone, its internal radio receives the signals from Boulder and displays the time. This unit also shows the indoor and outdoor temperature, which is sent by radio waves from its companion outdoor sensor (see Remote Thermometer in chapter 11).

How does Boulder set its watch? The Institute operates atomic clocks. These high-tech devices count the natural oscillations of an element such as cesium. Knowing how fast the oscillations occur, the atomic clock counts the oscillations and records the time elapsed.

INTERESTING FACTS
You can also hear the exact time being broadcast on short-wave radio (WWV radio) by tuning to 2.5, 5, 10, 15, and 20 MHz.

Audio Speakers

They transform electrical signals into the sweet sounds of music and the voices of talk radio and news.

An integral part of a system that plays music or make sounds, speakers can be found throughout the house. Most noticeable may be large speakers connected to the home stereo or entertainment systems.

HOW IT WORKS
Turn on a radio, or better yet a stereo system. Place your hand on the speaker to feel the vibrations. When a drum plays or a low-pitch sound occurs you can feel the speaker vibrate. If your sound system has a bass adjustment, turn the bass way up to feel the vibrations even more.

Most speakers have an electromagnet that vibrates as the radio signal changes. The electromagnet moves a diaphragm made of paper. As the diaphragm moves in and out, it moves air molecules at the speed of sound and creates sounds. Thus the speaker converts electrical signals into signals we can hear.

Phonograph (Record Player)

BEHAVIOR
Reads the etchings in vinyl disks (records) and converts them into electrical signals that can be amplified and heard as music and other sounds.

HABITAT
Increasingly difficult to find, phonograph players are found only in the homes of connoisseurs of music.

HOW IT WORKS
The most conspicuous parts of a phonograph player are the turntable and tone arm. The turntable is the circular platform that the record rests on. An electric motor underneath the turntable spins it at a constant speed.

The tone arm is the long, slender device that carries the cartridge and stylus (that picks up the signal from the record) at its free end. There is a weight at the other end of the tone arm to counterbalance the weight of the cartridge and to prevent the stylus from scratching the record.

The stylus is the piece of metal, diamond, or synthetic material that touches the record. It follows the spiral groove of the record as it spins, and bumps into the tiny swiggles in the record that represent the sounds. It transmits them to the cartridge, where they are converted into electrical signals that are amplified and fed to the speakers.

Of his hundreds of inventions, Thomas Edison was most proud of the phonograph. In 1877 he was trying to make a device to record telegraph messages automatically. Edison knew that sounds were vibrations and could vibrate materials, so he made a device that converted sound vibrations into etchings on a piece of foil. Besides inventing a new communications system, he kindled the race to invent other methods of recording sounds.

Video Cassette Recorder (VCR)

Transforms invisible bits of iron oxide on tape into *The Sound of Music* or *Top Gun*.

VCRs are co-located with television sets, often part of an even larger home entertainment system.

If you like to see how things work from the inside, try taking apart a discarded VCR. That will keep you entertained for hours.

There's far too much information in a video signal to store as magnetic bits in a straight line on a tape. Instead, the information is recorded at an angle to the tape, in a spiral pattern. To read the magnetic tape the "head" spins very quickly, 30 revolutions per second, past the tape. Meanwhile the tape is being pulled past the head at a more leisurely pace of a few inches per second.

The head, a magnetic detector, senses the tiny magnetic fields resulting from the pattern of the metal oxide coating on the film. It converts these magnetic fields into electrical signals that are played on the television screen.

Although VCRs have played a role in home entertainment for the last 30 years, they are quickly being displaced by newer technology, principally DVDs. Soon VCRs will be historic artifacts of the 20th century.

Fireplace

BEHAVIOR

It provides a place to hang Christmas stockings and a warm place to sit near on cold, winter evenings.

HABITAT

You can find fireplaces predominantly in living rooms, but houses that have one may have a second one (probably sharing the same chimney) in a family room or den. Some homes even have one in the master bedroom.

HOW IT WORKS

Traditional fireplaces are brick or steel enclosures for burning wood. However, today some burn natural gas instead of wood.

The firebox is where the combustion takes place. Fuel, oxygen, and heat are all required for fire. The chimney provides an avenue for smoke to escape. Warm air and combustion gasses rise and pull along surrounding air, creating an updraft. This pulls fresh air from the room into the firebox. The flow of fresh air sustains the fire, but also removes warm air from the room, making the fireplace an inefficient source of heat.

Many methods of improving the efficiency have been developed. Fireplaces can have vents in the brickworks surrounding the firebox so room air can circulate and pick up heat without being sucked up the chimney. Another way of circulating air to pick up heat is through openings in the grate that holds logs. Blowers attached to the openings can pull in room air and drive hot air back out to the room.

Many fireplaces have been converted to gas heat. A gas jet vents under a pile of fake logs and, when lit, appears to be a wood fire. You can distinguish a gas fireplace by the presence of a small metal pipe carrying gas and a shut-off valve outside of the firebox. Gas fireplaces also do not produce ashes.

Piezoelectric Lighter

BEHAVIOR
With the pull of a finger it generates a flame to light a fireplace, candle, or barbecue grill.

HABITAT
It can be found in the gadgets drawer in kitchens, on the fireplace hearth, or beside the barbecue grill.

HOW IT WORKS
Piezoelectric lighters use pressure to generate electric voltage. When you push the button or pull the trigger of a piezoelectric lighter you compress a spring. At the end of the pull or push a latch releases the spring to strike a piezoelectric crystal. You can hear the "whack."

These crystals have the unusual property of transforming mechanical energy—the whack—into electric energy. The result of whacking the crystal is that a large voltage builds up in the crystal, equivalent to the voltage discharged by a spark plug. The voltage can jump or spark across a small gap to ignite a gas fuel.

Lighters have a reservoir of butane gas in the handle. Before sparking the piezoelectric crystals, you release a stream of gas that catches fire when the spark occurs. This lets you light candles or paper in the fireplace.

Carbon Monoxide Detector

BEHAVIOR
Alerts you to the presence of carbon monoxide.

HABITAT
Found on walls and ceilings near appliances that have open flames, for example near fireplaces, gas water heaters, gas dryers, etc. Some recommend placement high in a room, but more seem to recommend placement near the floor.

HOW IT WORKS
Unlike smoke detectors that alarm as soon as they detect smoke, carbon monoxide alarms trigger when exposed to the gas over a length of time. There are three types of detectors and each works differently.

Metal oxide semiconductor models are plugged into a wall outlet. Electric current is required to heat tin oxide that reacts with carbon monoxide. High levels of carbon monoxide set off an alarm.

Another type of detector uses battery power rather than alternating current. The alarm sounds when a detector senses a change in the color of a gel-coated disc, which is caused by the presence of carbon monoxide.

Also battery powered, electrochemical alarms go off when carbon monoxide generates an electric current through a chemical reaction with platinum oxide.

Carbon monoxide is the leading cause of accidental poisoning deaths in America. Since the gas is invisible, odorless, and tasteless, detectors are the only way you will know that levels are unsafe. This gas is less dense than air, but mixes readily due to convection currents.

iPod

BEHAVIOR
Stores thousands of songs or hundreds of photos and videos and plays or displays these at your whim.

HABITAT
Found in pockets, backpacks, and pocketbooks until the end of the day, when they can be found on dressers.

HOW IT WORKS
The iPod is one of several portable digital media players. It stores songs, videos, or photos either on a hard drive or flash memory card. A hard drive stores data on flat, circular disks using tiny magnetic fields. Most computers store files on similar hard drives. Flash memory devices are a type of EEPROM (Electrically Erasable Programmable Read-Only Memory), a memory type that can be reused thousands of times and does not require electrical power to maintain its memory. Digital cameras use flash memory.

Users control iPods with a scroll wheel. They download songs and images from computers. Data is stored in one of several compression formats. Compression programs remove data that don't contribute to the quality of the sounds or pictures so they can be stored more efficiently. For example, sounds above human hearing levels can be removed without altering the perceived sound.

3 BEDROOM

ALTHOUGH THEY DO NOT HAVE as much visible technology as other rooms in the house, bedrooms have a collection of gizmos and gadgets not found elsewhere. So, just before you nod off to sleep, look around the room to spot these marvels of engineering.

Smoke Detector

BEHAVIOR

Detects smoke and alerts you to the possibility of a fire. More often than not, its ear-piercing alarm goes off as the result of not opening the fireplace flue or burning a casserole.

HABITAT

At least one should be mounted on the ceiling of each bedroom and at least one room on each level of a house. They should be at least six inches away from any walls. If mounted on a wall, the detector should be at least six inches away from the ceiling.

HOW IT WORKS

Although some smoke detectors use a light or laser (smoke interrupts the beam of light, setting off the alarm), most use an ionization detector. The detector holds a tiny piece of the radioactive element americium. The americium radiation ionizes gases in the air (predominantly nitrogen and oxygen), creating positive and negative ions. The ions permit a small current to flow between two charge plates, powered by a nine-volt battery. (This is the battery you should replace when Daylight Savings Time switches to Standard Time and back again.) Smoke particles attach to the ions and reduce the current flow between the two plates. Sensing a decreased current, the alarm switches on that horrible buzzer.

INTERESTING FACTS

Although almost every home has smoke detectors, these devices weren't around until just a few years ago. The first commercially available smoke detector for home use was invented in 1969.

Shades

BEHAVIOR
Allow you to cover and uncover a window with a single hand.

HABITAT
Found in a variety of rooms, adjacent to windows.

HOW IT WORKS
The shade is wrapped around a rod. As you pull the shade down, the rod rotates. Inside the rod a spring winds up. A catch mechanism at one end of the rod locks in position when you stop pulling downward. To get the shade up again, you pull sharply downward and this spins the catch outward (due to centrifugal acceleration), allowing the spring to hoist the shade.

The shade shown here does not have an internal spring to pull the shade up. Instead, you pull one side of the cord to raise it and the other side to lower it.

INTERESTING FACTS
The window shade operates like a car seat belt, but in reverse. The seat belt allows you to extend the belt with a gentle pull, but a sudden tug (an accident?) locks the catch mechanism, thus keeping you secure.

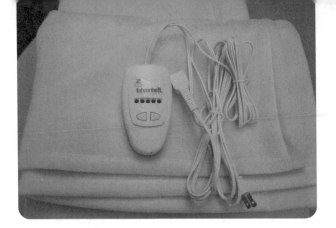

Electric Blanket

BEHAVIOR
Plug it in, turn it on, and you stay nice and toasty warm on a cold winter night. Too bad you can't take it on a camping trip.

HABITAT
Found on beds in colder climates and in closets of homes in moderate climates. When in use they plug into an electrical outlet.

HOW IT WORKS
The blanket converts electric energy into heat. Current passes through a circuit of electrically insulated heating wire that is embedded inside a blanket. The control keeps the blanket from overheating.

> The first electric blanket with thermostat went on the market in 1936 as a heating pad. It wasn't until the 1950s that they were called electric blankets.

Waterbed

BEHAVIOR
Supports the weight of a sleeping person on a layer of water.

HABITAT
Once found only in the bedrooms of the very hip, waterbeds can be found anywhere.

HOW IT WORKS
A waterbed is a plastic envelope that is filled with water. To prevent sloshing back and forth (Surf's up!) waterbeds have baffles inside. Baffles are internal walls that retard the movement of water—they stop the slosh. Some waterbeds also have air chambers to provide more rigid support along the edges and inside at several places.

Since a room temperature waterbed would draw heat away from your body quickly, waterbeds are heated. A heating pad is placed under the bed and a thermostat moderates the temperature (shown below).

Devotees claim that waterbeds provide better back support and are cleaner, since flakes of dead skin don't accumulate inside the mattress. However, waterbeds do require constant electric power (for the heater) and periodic addition of chemicals to kill algae and bacteria that can grow in the water.

INTERESTING FACTS
Charles Hall, a design student at San Francisco State University, invented the modern waterbed in 1968. It caught on big during the sexual revolution of the late 1960s and early '70s, but Hall was unable to get a patent because the concept had been previously described in several novels.

Digital Clock

BEHAVIOR
It flashes 12:00 until you reset it. Then, without power interruption, it shows the correct time in digital format.

HABITAT
Digital clocks can be found throughout the house. In the kitchen, the coffee maker, the microwave oven, the conventional oven, and other appliances have them. In the living room, den, or family room, the VCR and radio show digital time. The bedside clock is likely to be a digital clock. Answering machines and many other devices have digital clocks.

HOW IT WORKS
Digital clocks get their power from either a battery or the wall outlet. Battery driven clocks use a crystal oscillator to keep time. Clocks that plug into the wall count the number of cycles of the 60-cycle-per-second electric current, which is regulated at the electrical power generating plant. Most clocks don't have an internal power backup, so when they lose power they "forget" what time it is. Reconnecting them to power causes them to revert to the dreaded 12:00.

There are several methods for displaying the time. Some clocks have numbers painted on individual panels that rotate. Most digital clocks use seven-component LEDs (light emitting diodes) or a seven-segment LCD (liquid crystal display). With seven lights or bar segments you can create all 10 digits for display.

Clock Radio

BEHAVIOR
Wakes you up with the sweet sounds of easy listening music, or rattles your cage with heavy metal.

HABITAT
Usually positioned atop a nightstand in the bedroom. Located so the "Snooze" button is within easy reach while lying in bed.

HOW IT WORKS
Like telephone answering machines, the clock radio is a wonderful example of combining two existing devices to create something new. Since its introduction in 1947, the clock radio has been a bedroom mainstay.

The two devices operate separately, but can also function together. Older models have analog clocks and operate mechanically: gears rotate with the clock mechanism and trip a switch to turn off the radio (slumber mode) or turn it on (alarm mode). Newer models have digital displays and electronic circuits to control the alarm, various sounds, and the radio.

Emergency Radio

BEHAVIOR
Provides information (and music) from radio stations when other radios don't have power to operate.

HABITAT
Kept in closets or drawers until the "no-battery, no-power" emergency hits. Then it becomes the center of attention wherever the family is huddled together.

HOW IT WORKS
A nifty device that doesn't rely on the household electrical supply, this AM/FM radio can get its power from several sources. Solar cells in the top allow for solar power. But, since many situations where electrical power has been lost include bad weather, there is a hand-cranked dynamo on the side. Spinning the crank even a few rotations generates enough energy to power the radio for a few seconds. Power generated is stored in on-board, rechargeable batteries.

The radio also can be powered by disposable batteries and by household current. You can recharge the on-board batteries by plugging the radio into a wall outlet.

> You might own another dynamo or generator. Although not as popular as they once were, lights on bicycles are often powered by generators that rub against a tire. As the tire spins, it turns the generator and makes electricity to light the lamp.

Lava Lamp™

BEHAVIOR

Floating blobs in a glass sleeve move and change the pattern of light coming from the base. This action elicits "Oohs," "Ahs," and "Cool, man!"

HABITAT

Found in the bedrooms of the hip in the 1960s and '70s, these liquid motion lamps are enjoying a nostalgic comeback and can be found in the bedrooms or dens of baby boomers and Gen-Xers.

HOW IT WORKS

The moving blobs are wax that is heavier than the surrounding liquid when cool and lighter (less dense) when warmed. A blob sinks to the base where a 40-watt lightbulb heats it. It expands and, being less dense, floats slowly to the top, where it cools. Once cooled, it sinks back to the bottom where it is heated again.

Critical to making this work is picking two materials, the wax and surrounding liquid, that have about the same density. Companies that make the lamps keep secret the materials they use.

Edward Walker of England invented the Lava Lamp in the early 1960s, just in time for the youth culture to adopt it.

4 KITCHEN

KITCHENS ARE LABORATORIES where experiments in biology, chemistry, and physics are conducted every day. Nowhere else in the home can you find the diversity of tools and technology as are found in the kitchen.

The basic utility of a kitchen requires heat for cooking, cooling for food preservation, water for washing, plus electric power to run the wide assortment of appliances. And, in the midst of all the culinary activity, this is where people tend to congregate. The kitchen is a busy place.

Refrigerator

BEHAVIOR
Keeps food cold (4°C or 40°F), but not frozen, to slow down the growth of bacteria and keep the food fresh.

HABITAT
Most often found in kitchens, but some are also found in family rooms. One can be found in 99.5 percent of American homes.

HOW IT WORKS
Refrigerators take heat out of the insulated box (where you store food) and pass it into the surrounding room. So, opening the door of a refrigerator may blanket you in coolness, but because the refrigerator has to work harder to keep the inside cool, it is heating up the room.

A refrigerator's machinery pumps a fluid back and forth between the outside and inside. When the fluid is outside, it is hotter than the room so it gives off heat. When the fluid is inside the refrigerator, it is colder than the air there, so it absorbs heat.

Refrigerators have a pump or compressor for moving a low-boiling point refrigerant between coils on the inside and outside of the refrigerator. It is the pump or compressor that you hear start and stop. The compressor increases the pressure of vapor refrigerant, and this increases its temperature. The gas is passed through the coils in the back (or underneath) of the refrigerator where it cools in contact with the air in the room. It goes through a device to reduce its pressure before flowing into the tubes inside the refrigerator. Now cooled, the gas condenses into its liquid state, which lowers its temperature. The liquid is much colder than contents of the refrigerator and picks up heat from inside. The refrigerant cycles between being vaporized (so it can absorb heat) and condensed to a liquid so it gives up heat.

UNIQUE CHARACTERISTICS

Search for the ubiquitous dust bunnies that hide beneath and behind the refrigerator. Suck them out with a vacuum, not just for the joy of cleaning, but also to improve the operation of your refrigerator. The most common maintenance problem with refrigerators is dust coating, and thus insulating, the heat-exchanging coils. Keeping the coils clean allows the refrigerator to run more efficiently and saves you electric power and money.

The refrigeration process was discovered before electricity was available to power a refrigerator. The discovery was made in France in 1824, and 94 years later Kelvinator introduced the first refrigerator for home use.

Microwave Oven

BEHAVIOR
They bombard the water molecules inside food with radiation to heat it.

HABITAT
Usually found in kitchens, new models are small enough to fit comfortably in any room.

HOW IT WORKS
The oven uses electric power to generate microwave radiation with a vacuum tube called a "magnetron." A stirrer rotates above the cooking food to spread out the microwaves. It rotates about once a second and you can see it (or its shadow) slowly spinning when you open the oven door immediately after cooking. The radiation has a wavelength of about 5 inches (12 cm) and can pass through glass and plastic much like visible light can. This is why the glass cover on a food dish doesn't interfere with cooking the food.

Water, fats, and sugars, however, absorb the radiation. (Note: "radiation" is used to describe energy moving through space as either waves—electromagnetic radiation: light or radio waves—or particles—nuclear decay. A microwave uses the former type of radiation, not the latter.) The water, fat, and sugar molecules vibrate back and forth as they try to align themselves with the fast-changing electric field set up by the microwaves. This motion generates the heat that cooks the food.

UNIQUE CHARACTERISTICS

Look at the door of your microwave oven. Notice the latch that ensures the oven won't start until the door is shut. This ensures that the radiation, which can be harmful to living critters, doesn't escape. Check out the viewing window. Since microwaves can travel through glass, a glass window provides no protection from the radiation. The metal mesh on the inside of the glass, with openings much smaller than the wavelength of the radiation, blocks the escape of radiation but allows visible light through.

INTERESTING FACTS

Microwave ovens are energy efficient machines. In traditional ovens 50 percent or less of the heat generated warms the food, but microwave ovens operate at 70 to 80 percent efficiency.

Engineer Percy Spencer discovered the cooking effect of microwaves by accident. The story is that while experimenting with microwaves, he discovered that a candy bar in his pocket had melted. When he associated the melting with the microwave radiation, he first tried popping popcorn. When that worked, he made the world's first microwave mistake trying to cook a raw egg. It exploded, and Percy's mind exploded with the concept of microwave cooking.

Toaster

BEHAVIOR
Used for burning bread.

HABITAT
Found on most kitchen countertops.

HOW IT WORKS
Although there are lots of different styles of toasters, the most common is the pop-up toaster. It toasts bread by passing electric current through high resistance wire made of nickel and chromium. The resistance of the wire converts electric energy into heat, and as the wires heat up they radiate infrared (and visible) light that heats and toasts the bread.

After dropping a slice or two into the slots, you depress the lever. That starts the electric current flowing through the heating coils and sets a latch (mechanical or electromagnetic) to hold the bread down against the upward force of springs. It also starts a timer. The timer will shut off the current flow to the coils and release the latch so the toast can pop up. Some models use a bimetallic thermostat instead of a timer to control the process. Both are adjustable.

INTERESTING FACTS

Before the pop-up toaster, someone had to watch or anticipate when the bread would be toasted. As you can imagine, many a piece of bread was burnt to a crisp when the watcher got distracted by something other than the toast. Burnt toast frustrated Charles P. Strite to invent the pop-up toaster in 1921.

Toasting a piece of bread doesn't just make it darker; it fundamentally changes the bread. The toaster causes a chemical reaction, called a Maillard reaction, that generates the great smells and sweeter tastes. Heating causes the reaction to occur between amino acids and sugars in the bread.

Coffee Maker

BEHAVIOR
Transforms mere water and roasted beans into a stimulant that drives the nation.

HABITAT
Found on kitchen countertops everywhere.

HOW IT WORKS
There are several different types of these machines. The model shown here has you pour fresh water into a reservoir and add ground coffee beans into a container that is on top of the coffee pot. Turning on the power switches electricity through a heating element that curves around the inside of the base, below the coffee pot. The heating element is connected to a tube through which the water passes, but it only passes in one direction. The heating element heats the water in the tube, eventually boiling the water. Resulting bubbles rise up a tube, pushing hot water ahead into the container with the ground beans. The water drips down onto the ground beans and collects in the pot below. A filter retains the grounds while letting the liquid pass.

This method of brewing coffee is called the drip method. Other methods include the press and percolator. In a press, coffee grounds are flooded with boiling water and later filtered out. Percolators repeatedly send water through the grounds and back to the pot. A well in the bottom of the percolator pot collects water or coffee, heats it to a boil and sends it up the hollow tube. It spurts out the end, hitting the viewing cap, and drains down through the coffee grounds and falls into the pot below. Espresso is made by using pressure to force hot water through the grounds.

Passing hot water over ground coffee beans releases a host of chemicals (over 500 compounds), including caffeine. The story (or myth) of coffee's discovery is that a goat herder in Ethiopia noticed goats eating coffee beans. Seeing their unusual vitality, he tried the beans himself. From that simple beginning we now have coffee shops on every corner.

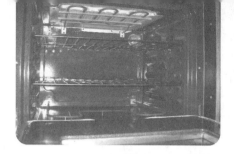

Conventional Oven

BEHAVIOR
It heats the air inside to cook food. Cookies, cakes, pies, roasts, and other delectables come popping hot out of an oven.

HABITAT
Conventional ovens are typically mounted in a cabinet in the kitchen.

HOW IT WORKS
Ovens operate in two modes: baking and broiling. In baking, heating elements, either gas flames or electric heating rods, heat the air inside the oven. A thermostat turns the heating element on and off to maintain a constant temperature.

A steak calls for a different form of cooking: broiling. Here a heating element in the top of the oven comes on and stays on until you shut it off, cooking the steak below by radiation. The heat radiates from the gas flame or electric heating element to the meat below.

INTERESTING FACTS
Many ovens today come with a self-cleaning feature. Cleaning an oven by hand is a real chore: reaching in to scrub off last week's baked lasagna is no one's idea of fun. Self-cleaning ovens have a high-temperature setting that burns off the spilt lasagna and cookie dough. The temperature is so high—900°F (480°C)—that it burns up all the spilt material. Accidentally opening the oven door during the cleaning process could cause burns, so the oven comes with a special door lock (that activates when the temperature inside reaches 600°) to keep it shut while the oven is at high temperatures.

Convection Oven

BEHAVIOR
This type of oven cooks food faster at lower temperatures than a conventional oven does.

HABITAT
In larger kitchens or the kitchens of serious bakers.

HOW IT WORKS
Convection ovens operate like conventional ovens with one difference: a fan forces the air inside to move. Heating elements (gas or electric) warm up the air inside. The fan blows air cooled by contact with the cooking food away so hotter air can come in contact. The constant circulation reduces cooking times by about 20 percent.

Most foods bake in an oven for a prescribed length of time. You can test cakes with a toothpick to see if the dough sticks (not cooked) or comes out clean. Turkeys often come with their own timer to let you know that it is cooked.

The turkey timer is a spring device made of plastic. The pop-out stick is held in place by a small dab of soft metal. When the turkey is cooked (inside temperature reaches 185°F), the metal melts, freeing the stick to move. A compressed spring pushes the now-free stick into the "cooked" position. After it cools, take a look at the timer and pull it apart.

Stove

BEHAVIOR
Supplies heat for cooking foods.

HABITAT
Usually the stove commands the central location of a kitchen or the central spot along the kitchen counters.

HOW IT WORKS
Stoves, traditionally heated by burning wood, are now almost exclusively heated with either natural gas or electricity. However, some stoves come not with visible burners but with a ceramic glass panel that uses radiant heat to cook.

Electric stoves generate heat by passing high electric current through a highly resistive metal, or burner. Burners are made of nickel-chromium metal, sheathed in stainless steel. The burner can heat up to become glowing or "red hot."

Gas stoves mix gas piped in from outside with oxygen from the air to burn. Older models require the gas stream to be lighted by hand. New models have built-in electric starters that "click" on when you turn the dial to "light."

The distinctive smell associated with natural gas is not natural gas, but mercaptan, a chemical added so you can detect the otherwise odorless gas.

Extractor Fan or Exhaust Fan

BEHAVIOR
Removes the smell of cooking liver and onions from the kitchen.

HABITAT
Found directly above or behind the stovetop.

HOW IT WORKS
In most systems a fan blows air into a tube that vents to the outside. This tube can run to the ceiling above the stove and out, or down to the floor and out.

Less costly to install are stand-alone fan systems. They blow air through a filter bed of activated charcoal to remove the particles and smells. Activated charcoal has a huge surface area that can absorb organic and other chemicals as they pass through the porous charcoal. Treating charcoal with steam activates it.

UNIQUE CHARACTERISTICS
The fan shown above hops up from the stovetop and sucks the air down to the basement below and then to the outside. When not in use, a motor pulls it back down into the stove cabinet.

Wine Cork Remover (Corkscrew)

BEHAVIOR
Permits access to the tantalizing liquid inside a corked bottle.

HABITAT
Usually resides in the utility drawer in the kitchen.

HOW IT WORKS
The model shown here is a popular design that is easy to use. You turn the top handle to move the screw into the cork. As the screw buries itself into the cork the fly handles rise. By pushing down on these handles you pull the cork out. This corkscrew provides two lessons in leverage, as well as opening bottles.

This is a great example of two machines in one gadget. First is the screw that you drive into the cork. Once in, it holds the cork very well, making it almost impossible to break the cork apart when pulling it out. Second is the rack and pinion gearing. The threaded handle acts as the rack and the two handles that fly out to the sides act as pinion gears. The handles give you great leverage to extract even the most recalcitrant cork.

Hand Mixer

Mixes ingredients into chocolate chip cookies and cakes of all persuasions.

It resides in one of the cabinets in the kitchen, alongside the other baking utensils.

An electric motor does double duty in the hand mixer: it both turns the beaters and turns the blades on a cooling fan.

Look at a hand mixer and wonder how one motor connects to two beaters and how the motor, with its drive shaft horizontal, spins two vertical beaters. A worm gear satisfies these conundrums.

The motor shaft supports a small fan and attaches to the worm gear. A worm gear has its grooves cut at an angle with its shaft. Worm gears greatly reduce the speed of rotation from the high speed electric motor to the slower speeds needed in mixing; and they also change the direction of rotation by 90 degrees.

The worm gear drives two counter-rotating gears attached to the sockets that hold the beaters. Having the beaters turn in opposite directions draws batter inward toward the beaters and then out.

Dishwasher

BEHAVIOR

Miracle of miracles: you put in last night's dirty dishes and in the morning you take out clean and sparkling dishes ready for breakfast.

HABITAT

Found in over half of the home kitchens in the United States, usually residing adjacent to the sink. Some are mobile; they hide under a kitchen counter until called on to do their magic. Then they are rolled to the sink to connect with the faucet.

HOW IT WORKS

If you squirt hot water laden with caustic chemicals hard enough, you'll get dishes clean. The water is hotter (130° to 150°F) and the chemicals (detergents) are stronger than you can stand for hand washing. A dishwasher is a waterproof container that heats and squirts water, releases chemicals at the right time, drains the water, and dries the dishes.

An electromechanical timer, just like in clothes washers, controls older models. Newer models come with a microprocessor that allows you to have a greater range of washing options.

Josephine Cochran is generally credited with inventing the practical dishwasher in 1885, and there are more than 30 other patents assigned to women for dishwashers at this time. Cochran was a relative of the inventor of the steamboat, John Fitch.

Trash Compactor

BEHAVIOR
Squeezes trash and garbage into a smaller space to fit inside trash cans and into garbage dumps.

HABITAT
Found in kitchens as a built-in appliance. Looks like a narrow dishwasher.

HOW IT WORKS
You drop trash into the bag inside the compactor. When ready to squish, you close the compactor and push a switch. The switch activates a motor in the base of the compactor. Through a chain drive, it drives two vertical screws that support a steel "ram." As the screws rotate, the ram is pulled down, compressing the cottage cheese container and chicken bones into a crumpled mess sure to frustrate future archaeologists.

When the ram has passed the "squished" point, it flips a switch that reverses the motor. The screws rotate in the opposite direction raising the ram.

Compacting trash defeats recycling and so may be frowned on in the future.

Garbage Disposal

BEHAVIOR
Grinds up and swallows garbage jammed down the kitchen drain.

HABITAT
The disposal itself is mounted directly beneath the kitchen sink drain. A wall-mounted switch that activates the disposal is usually located behind the sink.

HOW IT WORKS
Washing chunks of food down the drain is an invitation for Roto-Rooter to pay a visit. Garbage disposals chew up food scraps into sizes small enough not to lodge in the drainpipe.

A motor spins a cutting plate inside the disposal when the switch is activated. Movable cutters mounted on the top of the cutter plate slam outward due to centrifugal force. Food is caught between the spinning cutters and the stationary outer ring. During operation water runs through the disposal to pick up bits of chopped food and carry it through holes in the cutter plate, out the drain.

Food Processor

BEHAVIOR
It grinds, beats, smoothes, liquefies, whips, and purees foodstuffs.

HABITAT
Usually found either on the kitchen counter or inside a kitchen cabinet.

HOW IT WORKS
A heavy-duty motor drives the various blades and attachments at a variety of speeds.

Chester Beach and L. H. Hamilton patented the first electric food mixer and formed the Hamilton-Beach Company. But food processors are a relatively new addition to the kitchen scene. Carl Sontheimer worked on the design for a year before introducing the first food processor in the United States in 1973.

Bread Machine

BEHAVIOR
Emits wonderful odors and, after a two-hour wait, a loaf of warm bread.

HABITAT
Found on kitchen countertops and tucked away in kitchen cabinets.

HOW IT WORKS
When you think of all the different steps required to make a loaf of bread by hand, it's a wonder that a machine with only one motor and a heating element can duplicate the production. But it does.

Once you've input the ingredients and started the machine, its timer takes over. First it mixes the ingredients. Then, with the heating element on, it pauses to give time for the yeast to convert sugar into carbon dioxide and alcohol. The flour-water mixture becomes elastic enough to capture the carbon dioxide and form tiny bubbles throughout the bread. The motor kicks on to knead the dough, letting excess gas escape. Then the heating element comes on to bake the bread. After the programmed bake time has elapsed, the beeper tells you it's ready.

Bread machines are the greatest things since sliced bread, which, by the way, was invented in 1928. Otto Frederick Rohwedder worked for 16 years to perfect his bread-slicing machine. Before Otto's invention, we don't know what the greatest thing was.

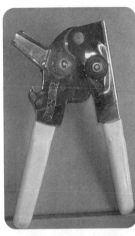

Can Opener

BEHAVIOR
Cuts through the lid of a metal can so you can spoon out the creamed corn or baked beans inside.

HABITAT
Electric models are usually found on kitchen countertops, plugged into an electrical outlet. Manual models are stored in drawers with all the other wonderful hand tools.

HOW IT WORKS
You lift the can of the tuna so the cutting wheel is positioned at the inside rim of the can. Pushing down on the lever drives the cutting wheel through the metal lid. Then, by hand or motor, you rotate the can so it passes under the cutting wheel.

INTERESTING FACTS
Consider this curious historical fact: can openers were invented 45 years after metal cans were invented. Does that mean that pantries held 45 years' worth of canned goods before the cans could be opened? No, people were ingenious enough to open cans without openers. (Witness a Boy Scout troop on a camping trip without an opener and you'll see innovation at work.)

The opener used today, employing a cutting wheel, was invented in 1870, 60 years after canned foods were introduced. The first electric can opener was introduced in 1931.

Water Filter

BEHAVIOR
It removes pollutants from water.

HABITAT
Filters can be installed on individual faucets or on the entire water system.

HOW IT WORKS
Even if the water leaving your local water plant is pure, it can pick up a variety of chemicals along the way to your faucet that you would be better off not consuming. A faucet filter removes these. The vast majority of water filters in homes use activated charcoal to remove impurities from drinking water.

Water passes through a bed of charcoal and the charcoal absorbs impurities. Key to the success of this filter is how fine the charcoal is ground. The grinding creates a huge surface area to catch the impurities. To increase the surface area further, charcoal is "activated" by heating it with various gases that open up tiny cavities in the charcoal.

UNIQUE CHARACTERISTICS
If you have one of these filters, try taking apart a used one. It's not easy. You can cut it open with a saw or break open the plastic shell with a hammer. Inside, you'll find the charcoal.

Water Cooler

BEHAVIOR
It delivers spring water to thirsty drinkers.

HABITAT
Water coolers are found on floor stands in kitchens. They are more prevalent in areas where the municipal water supply is of lower quality.

HOW IT WORKS
Once a week or so a delivery truck drops off a 40-pound, 5-gallon container of spring water. Consumers remove the plastic lid and place the jug upside down on the stand. Water is gravity-fed through a valve that lets water out and air in (so the jug doesn't form a vacuum).

UNIQUE CHARACTERISTICS
Older models require you to flip the 40-pound jugs upside down and position them in the cooler quickly so you don't spill (much) water. Newer models don't require opening before placing on the cooler. Instead, the seal is punctured as you lower the bottle in the cooler, which results in no spills.

Fire Extinguisher

BEHAVIOR
Hopefully never needed, it quickly puts out the flames of small kitchen or household fires.

HABITAT
Often stored in a kitchen cabinet or mounted on the wall in the pantry, utility room, or garage.

HOW IT WORKS
Most home extinguishers put out fires by covering the burning material and excluding oxygen (in the air). Without oxygen, the fire goes out.

Inside the extinguisher is the smothering agent and a pressurized container with carbon dioxide. Squeezing the handle pushes down on a rod that punctures the carbon dioxide container. The gas is released into the interior of the extinguisher. Squeezing the handle also opens a valve that lets the now pressurized agent escape through the nozzle.

The smothering agent in most extinguishers is a dry powder made of sodium bicarbonate, potassium bicarbonate, or monoammonium phosphate. Other extinguishers use carbon dioxide itself as the agent.

Home extinguishers contain only enough material to last less than a minute. They are suitable for a grease fire in a frying pan or similar small fires. Larger fires require the response of a fire department.

UNIQUE CHARACTERISTICS
Check the pressure gauge on your fire extinguishers. With time, pressure will be lost, making the extinguisher inoperable. Some can be recharged, but many of the lower-cost extinguishers must be replaced.

Cordless Phone

BEHAVIOR
Lets you gab without being tethered to one location.

HABITAT
These can be found anywhere a telephone is likely to be found, but are especially useful in kitchens where the cook can stir and talk at the same time.

HOW IT WORKS
A cordless phone is a combination of a telephone and two FM (frequency modulated) radio stations. The base receives calls the same way a traditional phone does. Then it broadcasts the voice as an FM signal to the hand unit.

The hand unit receives the FM signal, converts the radio waves into electric signals, and plays them through a speaker. It also converts the phone user's voice into electric signals and then an FM radio signal that travels back to the base. There the radio signal is converted back into an electric signal and sent out along the telephone wires.

The two FM radios operate at slightly different frequencies so the two voices don't interfere with each other. In normal radio operations, like on a CB (citizens band radio), both voices are carried on one frequency, which requires that each user signify when he or she is finished talking by saying "over."

Automatic Dustpan

Sliding the lever to one side starts the central vacuum cleaner (see chapter 2). Sweep the fallen Cheerios and assorted dirt into the dustpan and watch it disappear.

You'll find this in kitchens and possibly other rooms with hard floors.

The slide turns on the central vacuum system, which creates the suction to pull dust and dirt away. To see the Cheerios again, go to the central vacuum collector (see chapter 7), which is probably located in the garage.

5 BATHROOM

BATHROOMS ARE FUNDAMENTALLY different from all the other rooms in a house. Here, the presence of water and water vapor requires special coverings on walls and floors. Exhaust fans are needed to remove excess water vapor. And, with water and electrical appliances close together, special switches are needed to prevent electrical shock. We've grown up with plumbing and overlook its complexity and importance, until a toilet clogs or a faucet leaks.

The water system in U.S. homes today has and continues to evolve, as better ways are devised to provide water, remove wastewater, and protect people. The system is composed of four sets of pipes and many fixtures or appliances. Fresh water comes from a water main buried under the street. You can follow the water main by noting where gate valves are located in the street. Look opposite fire hydrants and at roads entering housing developments and you'll find the cover of a gate valve flush with the street. These valves allow utility workers to shut off part of the system without having to shut down the entire water supply.

From the water main, water passes through a shut-off valve and into a water meter. The meter measures water usage and reports the usage to meter readers driving by in trucks (many now have radio transmitters).

Incoming water is pressurized (usually by gravity—flowing from water tanks on a nearby hill or water tower) to flow through the pipes in your home. If your house is located atop a hill, you may have lower pressure than your neighbors who reside at lower elevations.

Once in your house, the incoming water feeds two systems of pipes. The first carries cold water (or unheated water) to sinks, showers, tubs, refrigerators, washers, and other appliances. One of these pipes carries water to the hot water heater, where once heated the water is carried in a second set of pipes to many, but not all, of the places the cold water goes. These two sets of pipes are the supply side.

Every drop that comes into your house has to leave and, except for that tiny percentage that evaporates, has to drain into the municipal sewer line (or septic tank and field). Unlike the supply side that is pressurized, water leaving drains out by gravity. So drainpipes feed into a "soil stack" that carries the waste into the house drain, which dumps into the sewer line. The sewer line also operates under gravity, which means that the pipes must be oriented downward towards the sewage treatment plant. If the plant is miles away, the pipes would be so far underground as to be impractical, so lift stations along the way pump the waste up and redeposit it into other downward-sloping lines.

One more set of pipes completes the picture. These are pipes that carry air and allow air pressure in the drainpipes to equalize, preventing vacuum blockage in the drainpipes.

This overview covers houses connected to municipal water and wastewater systems. Some rural homes are not tied into either the municipal water supply or sewer lines. They draw water from home wells and distribute wastewater into septic tanks and drain fields under their yards.

Ground Fault Interrupt (GFI) Switch

BEHAVIOR

It automatically interrupts electrical power when it senses a surge in current.

HABITAT

Sometimes called ground fault circuit inter-rupters (GFCI), these switches are most often found adjacent to sinks in bathrooms. They can also be found at other locations where people might be using electrical appliances near water: near swimming pools or hot tubs, in garages, and near wet bars.

HOW IT WORKS

The GFI switch shuts the electric flow in cases where you might be the electrical conductor between an appliance and ground. How could that occur? If you use your electrical razor in the bathroom and it is not grounded, electricity could flow from the razor through you to the cold-water pipe when you touch the pipe. The pipe acts as a ground, and if you are holding a hot wire or faulty appliance with one hand and a ground with the other, you will be shocked. The GFI switch senses a sudden surge in current and shuts off. Two coils inside the switch sense the flow of current through the hot terminal and the neutral terminal. If there is an imbalance between their current flow, it shuts the circuit within a few milliseconds.

INTERESTING FACTS

Some industry experts estimate that as many as 200 electrocutions a year could be prevented if everyone used GFI switches when operating electrical appliances near cold-water pipes.

Hair Dryer

BEHAVIOR
Blows hot air to dry your hair or warms you up on a cold morning.

HABITAT
Usually found in one of those cluttered drawers in the bathroom, or possibly left out on the countertop.

HOW IT WORKS
Hair dryers combine four devices to provide an instantaneous blast of hot air: a heating coil, a fan, a thermostat, and a switch.

The heating coil is made of wire composed of nickel and chromium, as is the heating wire inside your toaster. It converts electric energy into heat quickly. The fan is a small motor attached to plastic blades that draws air in from outside and blows it past the heating coil. Air spends only a fraction of a second within the dryer before it has been heated enough to do its job. The thermostat ensures that the dryer doesn't overheat.

New hair dryers come with a ground fault interrupter (GFI) switch that turns it off if you should drop it in the bathtub full of water. (See GFI switches for more information on how these work.)

The hair dryer works by blowing warm, dry air over your hair. Warm air can hold more moisture than does cooler air, so water is evaporated more quickly by warm air. The blower keeps a stream of dry air coming and pushes away air that has already picked up moisture.

Electric Toothbrush

BEHAVIOR

It keeps your pearly white teeth pearly white and hygienic.

HABITAT

Found on bathroom counters, plugged into wall receptacles.

HOW IT WORKS

A small DC electric motor inside the brush drives the bristles back and forth. What's especially cool is how the battery inside the brush is recharged. You can imagine that you don't want to use an electric appliance, plugged into the 120-volt outlet, in your mouth with water running. The shock potential is too high. So rechargeable batteries inside the handle power the toothbrushes. Looking at an electric toothbrush, you see that there are no metal contacts on the handle or inside the recharging station. How then does the battery get recharged?

Inside the recharger are the primary coils of a transformer. The other part (secondary coils) of the transformer is in the handle. The recharging unit creates a magnetic field in the primary coils from the electric field of alternating current (household power). The secondary coils of the transformer, inside the handle, convert this magnetic field into an electric field that powers the battery.

Traditional electric toothbrushes vibrate about 5,000 strokes per minute, while newer "sonic" toothbrushes vibrate seven to eight times faster. The faster vibrations energize saliva in your mouth to dislodge food particles. Electric flossing machines vibrate about 10,000 times per minute to clean between teeth.

A third type of electric toothbrush, the Waterpik™, squirts a tiny stream of water on and between teeth to remove particles.

INTERESTING FACTS
Studies have shown that electric toothbrushes are slightly (7 percent) more effective than manual brushing. Originally, they were intended to help people who have limited mobility to brush their teeth, but now these devices have become mainstream appliances.

Faucet

It is the on/off switch and volume control for water. In some cases, it also controls the temperature of the water. It is a valve for controlling the flow of water.

Found at every sink inside the house, and along outside walls where water might be needed.

In its simplest form the faucet handle screws down to seat a rubber or plastic washer over an opening in a water pipe. The screw provides the mechanical advantage needed to stem the tremendous force exerted by water under pressure.

More complex faucets allow one-handed adjustment of the water temperature. The valve controls the flow of water from both hot and cold water pipes. As you rotate or shift the handle, it changes the percentage of water coming in from the two pipes. The water enters a "mixer" before exiting the faucet so you get water of a uniform temperature.

Sink or Basin

BEHAVIOR
It provides a venue for mixing water, soap, and various potions used for cleansing and generally making one's self more beautiful or at least more hygienic. In the kitchen, sinks are used to wash carrots, tomatoes, and dirty dishes.

HABITAT
Kitchens, bathrooms, and utility rooms are the usual locations.

HOW IT WORKS
Most sinks are bowls made of ceramics or porcelain (bathrooms) or stainless steel (kitchens). They have one or two faucets to deliver hot and cold water. Public sinks now come equipped with automatic, infrared-sensing valves to turn the flow of water on and off. Water drains out of the sink into a pipe that runs to the sewer line.

INTERESTING FACTS
Although people have been using bowls or basins to capture water for washing for centuries, the modern sink is relatively recent. The sinks we have today weren't available a century ago.

Plunger

BEHAVIOR
Creates suction strong enough to dislodge clogs in the plumbing.

HABITAT
This device is usually kept out of sight and out of mind. If exposed, it appears behind the toilet, ready to work but out of the way.

HOW IT WORKS
If you've never used one, you are indeed lucky. The plunger—its official (but never uttered) name is the "hydroforce blast cup"—is used to move or break up stuff clogging the pipes leading to the sewer. You press down, creating some pressure in the pipes, but more importantly creating suction. By vigorously pulling up, the suction dislodges and breaks up the offending material.

Clean plungers make great toys. Jam one onto a smooth floor and whack the handle. Stick two together, mouth-to-mouth, and challenge young children to pull them apart. Stick them on a wall for an instant banner holder.

Toilet

BEHAVIOR
The sanitary disposal system for biological byproducts.

HABITAT
The anchor client in bathrooms. In other countries toilets are located in "water closets" and bathrooms are for reserved for taking baths. In the United States, bathrooms combine both functions and appliances.

HOW IT WORKS
Think you know? It's not as simple as you may think. The toilet is a marvel of engineering, flush with the success of countless innovators stretching back to Thomas Crapper in 1886.

Lift the lid to the reservoir to peak at the inner workings. Turning the handle lifts the flapper valve, which, once lifted, stays in place until all the water has drained out. As the water level drops, the ball float lowers, which turns on a valve connected to the end of the float arm. The valve lets fresh water come in from your cold water pipes.

There is no valve separating the stuff in the toilet bowl from the sewer line. (Hence those weird, but possible, stories about rats, snakes, and who knows what else crawling up a sewer line and into a toilet.) On flushing, the out-rushing water pulls the bowl contents up and over the sill in the bowl. Once over the sill, it falls/flows down the sewer pipe to the wastewater treatment plant.

Once the reservoir water has emptied, the flapper falls shut and the in-rushing water refills the tank. As the water level rises, the ball float rises. When sufficiently high, the float arm shuts the valve letting in water. That's the theory. My toilet continues to run, even after I fiddle with it repeatedly.

INTERESTING FACTS
Although we honor Thomas Crapper as the inventor, the wash-down toilet with ball float—the machine in most bathrooms today—was someone else's improvement of his more basic design.

> Modern toilet paper predates toilets. Although toilet paper has been used since at least A.D. 590, modern toilet paper was invented in 1857— 31 years before the flush toilet.

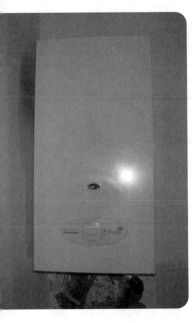

Tankless Hot Water Heater

BEHAVIOR

It provides hot water on demand, rather than storing it.

HABITAT

It is found very close to where the hot water is needed. A traditional hot water heater may source water for sinks, showers, and washers throughout a house over great distances. If the pipes are exposed to outside or ground temperatures, the water can lose heat along the way. A tankless hot water heater provides water for a local use. Since the water is hot instantly, there is no need to run water while waiting for the hot water to arrive.

HOW IT WORKS

Like traditional hot water heaters, tankless heaters use either electricity or gas to heat the water. To heat it quickly, the water pipe runs through a radiator adjacent to the heating element. The hot water heater turns on when you open the hot water faucet.

INTERESTING FACTS

Tankless water heaters make a lot of sense depending on how you use water and how your house is plumbed. If the pipes travel long distances to get from the traditional tank hot water heater to where you need the water, you lose much energy. And, if you leave home for days at a time, a traditional heater keeps using energy while a tankless heater doesn't.

Shower

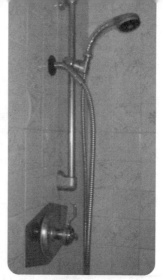

BEHAVIOR

Although being drenched with water is generally considered a bad thing, this is one place where it is eagerly anticipated. It is comparable to standing under a very warm waterfall.

HABITAT

Found in one or more of the bathrooms of the house. Typically, an exhaust fan is located adjacent to the shower, mounted either in the ceiling or wall, to remove vapor-laden air.

HOW IT WORKS

Two pipes supply water, one from the hot water heater and the other from the outside line (cold water). One or two manual valves control the ratio of the two streams, and they enter a "mixer" where they mix to come to an intermediate water temperature. The mixed water flows to a "head" that sprays out water through many small openings. The cascading warm water hits your tired body and evokes "ahs."

The water hitting the floor flows toward and into a drain that connects to the sewer. Blocking the opening is a drain cover or grate that collects soap scum, body hair, and other stuff that no one wants to clean up.

Plumbers install cut-off valves ("service valves") upstream of the shower fixture. This allows repairs to be done on the fixtures without having to shut off the water to the entire house.

Does your shower have a shower curtain? Have you noticed that the curtain is sucked inward when the shower is running? The cascading water entrains air molecules, pulling them along and away from the sides. This creates lower air pressure that sucks the curtain in.

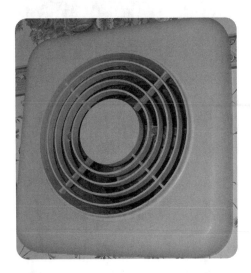

Exhaust Fan

BEHAVIOR

Removes high-humidity air from the bathroom and pushes it to the out-doors. Some operate automatically, but most require someone to turn them on. These are used to reduce mildew in the bathroom and to make the air feel more comfortable.

HABITAT

Found in the ceiling where warm, moist air collects, or high along an outside wall.

HOW IT WORKS

The fan blows air out of the bathroom, either into a vent pipe or directly to the outside. It pulls air from inside the bathroom and pushes it outward.

As the motor spins the fan blades, the blades deflect air toward the exhaust opening.

Bathtub

BEHAVIOR
Holds lots of water for a person to soak in.

HABITAT
Found almost exclusively in large bathrooms. In many homes only one bathroom comes equipped with a bathtub. If you own a rubber duck, this is where it will be found.

HOW IT WORKS
Tubs are made of acrylic or fiberglass that can be molded into a variety of shapes and sizes. Older tubs were made of porcelain-covered steel or galvanized steel.

Tubs have hot and cold water that the bather mixes to get the preferred temperature. Bath bubbles are added as an option.

Bathing is an ancient custom that seems to have been forgotten during the Dark Ages and the modern age. Benjamin Franklin is reputed to be the first person to import a bathtub to America. And, although the bathtub is a restful place, more people die in bathtubs each year than die from lightning strikes. More people die in bathtubs than from shark attacks, too. Of course, not too many sharks get into bathtubs.

P-Trap

BEHAVIOR
Prevents sewer gases from entering the home through sinks and drains.

HABITAT
Beneath every drain in the house is a P-trap.

HOW IT WORKS
The P-trap, which looks more like the letter U, traps wastewater in the bottom of the U. As wastewater flows out of a sink, it displaces the water that had been occupying the trap. The trap keeps water in the U to prevent gases (generated by bacteria in wastes flushed down the drain) from coming up through the sink and into the room.

At the bottom of most P-traps is a large nut or cleanout plug. When the P-trap becomes clogged you can clean it out (what a fun job this is!) by opening the cleanout plug with a large wrench. Make sure you have a bucket under the plug to catch the standing water from the trap.

Drain parts tend to come in three colors: white (PVC plastic), black (ABS plastic), and silver (metal).

Bathroom Scale

BEHAVIOR
Causes anguish among those on diets as they read the scale in disbelief. It measures the body weight of the person standing on top.

HABITAT
Usually found on the bathroom floor.

HOW IT WORKS
It's an amazing piece of engineering. When you stand on a mechanical scale, you depress its upper surface a very short distance, but move the scale in proportion to your weight. Four levers support your weight and transfer a small part of it (about 8 percent) to pull against the main spring. As the spring moves it pulls on a set of gears, like the rack and pinion gears that provide steering in your car. The pinion gear rotates the dial face to indicate your weight.

Some scales provide digital readouts. There are different ways these can work. One is that the levers depress a strain gauge that changes resistance as more weight is added. The resistance is measured, translated into poundage, and displayed on an LED display.

INTERESTING FACTS
Traditional mechanical scales have springs that are stretched fully by a weight of only 20 pounds. The levers reduce the weight they have to hold so larger and more expensive springs aren't needed.

6 HOME OFFICE

A GENERATION AGO a home office would have only a tiny fraction of the capabilities of an office today. Then, a manual or electric typewriter would signify that someone does serious work at home.

The miniaturization of electronics, the switch from analog to digital communications and data handling, and the proliferation of new equipment have fundamentally changed the home office. Now people may have better, faster computers and peripherals at home than at their company offices. And, whereas before people might have taken personal papers to the office for copying or faxing, now they can do it at home. The switch from film-based photography to digital has added more peripheral equipment to the computer that allows home production of graphics from scrapbooks to posters.

Surge Protector

BEHAVIOR

This device distributes electric power to several appliances and protects them (and you) from sudden surges in electricity.

HABITAT

Found around most computers and home entertainment systems. One end is plugged into a wall outlet and the other connects to one or more devices needing electric power.

HOW IT WORKS

Inside the surge protector are several components that divert excess electric power (above 120V) to a ground wire. There is also a fuse in case the diverted circuit doesn't work properly. Some protectors also have components to average out small fluctuations in the voltage.

One way of diverting excess current uses semiconductors that have variable resistance. In normal operation they present such a high resistance that electric power doesn't pass through them. But when a surge occurs, the higher voltages pass through the variable resistors to a ground. The device that performs this voltage switching resistance is a metal oxide varistor, or MOV. MOV surge protectors can be damaged by surges such as lightning strikes.

Minor fluctuations in the voltage are suppressed with a "choke," or electromagnet. Current passing through the electromagnet sets up a magnetic field that resists changes. As the current changes, the magnetic field induces an equal and opposite change in the current, thus balancing or at least reducing the fluctuations.

Here's some trivia to show up your geek friends: what's the different between a voltage surge and spike? A surge lasts at least three nanoseconds—3 millionths of a second—while a spike lasts two nanoseconds.

Uninterruptible Power Supply (UPS)

BEHAVIOR
Provides a steady supply of power through spikes, surges, and short power outages.

HABITAT
Found plugged into a wall outlet and connected to computers and other sensitive electronics equipment.

HOW IT WORKS
A UPS stores electrical energy in a lead acid battery like your car battery. Current from the wall outlet (alternating current) constantly charges the battery through a power supply. The battery provides direct current to an inverter that makes alternating current from direct current, and this alternating current supplies the computer with the power it needs.

Most home office UPS systems operate on a standby mode. They provide alternating current from the wall outlet to the computer unless there is a power outage. Within a few milliseconds of an outage, power begins to flow from the battery to the computer.

In more expensive systems (continuous UPS), the computer always gets its power from the battery. When a power outage occurs, the computer experiences no interruptions since it was already drawing power from the battery.

INTERESTING FACTS
Having a UPS can save wear and tear on your computer's hard drive. The battery provides time to safely shut down the computer rather than having the power cut off abruptly.

Power Supply

BEHAVIOR

Supplies voltages other than 120V alternating current (AC) to a variety of devices from printers to modems to answering machines. If you recharge your cell phone, laptop, or dust buster, you plug a power supply, sometimes called a wall wart, into an electric outlet.

HABITAT

They hang out at electrical outlets everywhere.

HOW IT WORKS

Look at the front and back sides of the power supply. It shows what it expects for an input voltage (120V AC in the United States or 240V AC for many other countries) and what it outputs. The output can be AC, in which case the device is just a transformer. It changes the voltage from the house voltage to what the appliance needs to operate.

If the output is direct current (DC), the power supply transforms the voltage to the appropriate voltage and rectifies the current or changes it from alternating to direct.

Inside a transformer are two coils of wire, each wrapped on opposite sides of an iron core. One set of windings connects to the house voltage and the other connects to the appliance. The two sides will have a different number of windings around the core. The ratio between the number of windings determines the voltage output: fewer windings on the appliance side give lower voltage.

UNIQUE CHARACTERISTICS

Even when you're not charging your cell phone, if its power supply is plugged in, it is using energy. Touch a plugged-in power supply—it will feel warm. The warmth tells you that energy is being wasted. Over the course of a year the waste is enough for a nice dinner for two.

Telephone

BEHAVIOR

It rings at dinnertime, usually prompted by a telemarketer. It can ring at other times of the day and it allows you to talk to almost anyone anywhere in the world.

HABITAT

Most homes (about 94 percent in the United States) have at least one phone, usually located in the office, kitchen, hallway, or other central place. In some homes, there's one in each room.

HOW IT WORKS

Two wires come into your house through the house protector, which is attached to an outside wall. One wire is called the ring and the other the tip. These connect through the hook switch to the handset. You pick up the phone and that closes the hook switch that connects you to the circuit. The microphone in the lower part of the handset converts the sound of your voice into electrical signals that travel to Aunt Millie. The receiver converts the electrical signals coming from Aunt Millie's phone back into sound.

An electronic device keeps you from hearing your own voice. A bell inside the phone alerts you to an incoming call. The touch pad is an array of switches that turn on pairs of sound generators. Those sounds are converted to electrical signals that guide your call through the many switches between your house and Aunt Millie's.

Necessity, they say, is the mother of invention. An undertaker who was convinced that his competitor's wife, who worked as a telephone operator, was steering his business to her husband invented the rotary dial. He created a system for removing operators, with their human foibles, from the communications loop.

Telephone Jack
(RJ-11 Connector)

BEHAVIOR
Provides a quick connection for a phone or modem.

HABITAT
Found in many of the rooms of a house, usually above the baseboards, mounted in a wall.

HOW IT WORKS
A "jack" is a socket that the telephone plug fits into. The RJ stands for "registered jack," meaning that it is a standard connector. Most houses have four telephone wires, although only two are needed per phone line. A second pair of wires typically is installed in case a second phone line is added.

The jack connects the two center wires, usually red and green, coming from the house protector to the telephone. The other two wires, usually yellow and black, can be accessed with a small device that plugs into the jack and switches the connections so the black and yellow wires connect to the center pins.

A jack that provides service for only one phone is an RJ-11. If it can be used for either of two lines, the jack is an RJ-14. There are many other types of registered jacks. See http://arcelect.com/RJ_Jack_Glossary.htm for the list.

A small plastic tab locks the plug in the jack in place so it doesn't pull out easily. You have to lift the tab to remove the plug.

Answering Machine

It intercepts phone calls, identifies you, explains that you are really doing something important at the moment, and invites callers to leave a message. You can listen to the recorded messages at your convenience.

HABITAT
It resides next to the telephone or is part of the telephone.

HOW IT WORKS
Like many creative inventions, the answering machine resulted from combining two existing devices: the tape recorder and telephone.

Older models record both outgoing and incoming messages on small cassettes of magnetic tape. Voices are recorded on the tape with a write head that aligns the magnetic fields of the metal oxides embedded in the tape. A read head interprets the alignment of the magnetic fields and converts it into electric signals that travel to a speaker where the electric signals convert to acoustic signals.

Newer models use digital recording. The voice is converted to an analog electric signal. An analog to digital converter transforms the continuously varying signal into a series of binary bits that are recorded in a random access memory, much like the one in your computer.

The earliest telephone answering machine was invented in the 1930s. However, until the 1970s, telephone answering machines were not commonly found in American homes.

Computer

BEHAVIOR
Sits atop its throne, prime mover of the home office. It processes bits of data at amazing speeds and enables a wide array of peripheral devices from digital cameras to personal robots.

HABITAT
The computer usually occupies the high-value real estate in the home office, sitting front and center on the desk. However, some are relegated to sit on the floor and control the work through cables, with only the monitor on the desk.

HOW IT WORKS
Home computer systems are composed of several major components (hardware): a central processing unit (CPU), input/output devices (keyboard, mouse, printer, monitor), memory (random access memory and hard drives), a clock, a power supply, and wires that connect

the components. Software, or computer programs, provide directions for processing data.

Computers do very simple operations but do them so quickly that complex functions can be achieved much faster than if humans were attempting to do them. As hardware speeds have increased (approximately doubling every two years, roughly following Moore's Law, see below), software can be more complex and lengthy allowing more functions and greater details.

INTERESTING FACTS

When personal computers first appeared there wasn't much you could do with them. There were some limited word processing programs and games. Quickly people created more things to do with computers, and engineers made faster and more powerful computers that spurred software and hardware developments. Today companies and individuals are still racing to find the next "killer application" and the next improvement in hardware.

> Moore's Law states that the number of transistors on integrated circuits will double every two years. Gordon Moore made this prediction in 1965 and, surprisingly, it has been accurate.

Mouse

BEHAVIOR
It allows you to move the cursor on a computer monitor and to select words, images, or areas of the display for capture and processing. It also allows you to access some other commands.

HABITAT
It lives on a mouse pad adjacent to the computer keyboard.

HOW IT WORKS
Mechanical mice use a ball that rolls across the mouse pad. As it moves, it rotates two perpendicularly mounted wheels inside. As each wheel spins it turns another wheel with slits. An infrared light emitting diode (LED) shines through the slits and infrared photo diodes count the number of slits that pass. Each passing slit signifies movement in either of the two directions.

Optical mice take pictures of the underlying mouse pad 1,500 times a second. Changes in the image indicate motion. An LED shines down on the mouse pad and photo diodes record levels of reflected light that indicate the position. Because they use light to detect motion, optical mice can act squirrelly on highly reflective surfaces or on transparent surfaces, such as glass tabletops. In these situations, mechanical mice perform better.

UNIQUE CHARACTERISTICS
Grab an old mouse on its way to the landfill and take it apart. It's an easy dissection; only a few screws hold it together. Looking inside, you can understand how it works.

INTERESTING FACTS
You might call it a "mouse," but the original patent called it "an x-y position indicator for a display system." Douglas Engelbart of Stanford Research Institute invented the mouse in 1963.

Keyboard

BEHAVIOR
Allows entry of numbers, letters, and commands into the computer.

HABITAT
Keyboards are often stored either on the office desk or on a tray that slides under the desktop.

HOW IT WORKS
Keyboards are collections of switches arranged in a common format so you can go almost anywhere and use a computer.

Each key is a single switch. Pressing down on one closes its switch, allowing electric current to flow. A microprocessor (in essence a limited function computer) interprets the signals and looks up in a memory bank what the signals represent. Pushing the Shift key and the letter A represents something different from pushing the Control A, and the processor figures out which symbol or action is required.

The cable connecting the keyboard to the computer carries electrical power to the keyboard and the input data from the keyboard to the computer. Wireless keyboards use infrared or radio communications to get information to the computer and require an onboard battery for power.

UNIQUE CHARACTERISTICS
Look out for old keyboards, as they are great to take apart. You will be impressed at how simple a keyboard is. And with the keys you take out, you can spell messages to glue to your refrigerator.

Modem

BEHAVIOR

Transforms the electronic bits of data into a form that can be carried over telephone wires or cable, or sent by a wireless transmitter.

HABITAT

Modems can be found adjacent to computers on desktops. They plug into computers and into phone lines or cable lines, plus electrical power outlets.

HOW IT WORKS

This machine modulates and demodulates (changes) a carrier signal so the resulting signal carries information that can be understood at the destination.

Dial-up modems allow computers to transfer information (slowly) over telephone wires at voice frequencies. Faster is the asymmetric digital subscriber line (ADSL). The "asymmetric" refers to the differences in the rate of speed in uploading and download data: it's much faster downloading than uploading. This asymmetry results from a much broader bandwidth being allocated for downloading than uploading. You can use the same telephone wires simultaneously for talking and surfing the Web as voice communications occur in a different part of the frequency spectrum than is used for ADSL operations. To use ADSL you must be within about three miles of a telephone central office.

Cable modems encode signals at radio frequencies to send and receive over the television cable system. The same space (6 MHz) is allocated on a cable system for Internet downloads as for Comedy Central, or ESPN. The 2 MHz bandwidth is allocated for uploading information. Some satellite systems allow you to download from satellite, but require you to use a dial-up line to upload.

Facsimile Machine (Fax)

BEHAVIOR
It converts printed documents into electric signals that can be transmitted through telephone wires to distant machines.

HABITAT
Found in many home offices, faxes are usually located near computers.

HOW IT WORKS
Faxes can be stand-alone machines or scanners that rely on computers to send their output. In either case, the critical component of the machine is its scanner.

The scanner has over 1,700 sensors (photodiodes) to cover the width of a standard piece of paper (8.5 inches). Each one gives a reading of reflected light levels over 1,000 times down (the length of) the piece of paper. This sampling results in almost two million bits of data per page. To save time sending and receiving this data, it is compressed using one of several different schemes. Light for the photodiodes to sense comes from a fluorescent bulb. Either the paper is pulled past the bulb and sensors, or the sensors and bulb are pulled over the top of the paper.

Receiving an incoming fax, the machine decodes the signal and displays the image as a graphics image, which you can send to the printer.

Stand-alone models connect directly to a phone line that you can see trailing out of the back of the machine along with an electrical power cord. Scanners connect to a computer and not to a phone line. The computer can be connected directly to a phone line or through a DSL or cable modem.

Crude as it might have been, the first facsimile machine was invented in 1843 by Alexander Bain. It worked with the telegraph system and never caught on. Faxes became commonplace only after Dr. Alexander Korn developed photoelectric scanners.

Monitor

BEHAVIOR
Lets you see images and documents that are stored in the computer.

HABITAT
Accompanies a computer, usually less than three feet away.

HOW IT WORKS

Most monitors are either cathode ray tubes (similar to television sets) or liquid crystal displays (found in laptops and some desktop monitors).

A cathode ray tube is an evacuated tube. An electron gun shoots a stream of electrons at the fluorescent screen. The beam of electrons is guided by magnetic or electric fields so it can sweep across the screen and down to form a picture. Traditional television sets work this way.

A liquid crystal display (LCD) starts with a stream of light passing through polarizing filters that sandwich a liquid crystal. Each picture element (pixel) is a cluster of liquid crystal molecules connected to two electrodes. The electrodes for each liquid crystal carry electrical charge to it, causing it to change position and change the level of light it allows past, which feeds the picture signal.

In comparing cathode ray tube models to LCDs, note that each is measured differently (although both are measured diagonally across the screen). The advertised size for cathode ray models includes the plastic case in addition to the screen. Besides clarity of image and depth or number of colors, what is important is the size of the viewable screen.

Philo Farnsworth originally conceived the cathode ray tube. Unlike the majority of inventors in the field of electronics, Farnsworth enjoyed few resources and limited education. Raised on a farm in Idaho, Farnsworth raced and beat the leading companies of the era to develop television.

Inkjet Printer

BEHAVIOR
It spits tiny drops of ink onto paper so you can share great ideas, photographs, and business plans with the world.

HABITAT
Found in many home offices, adjacent to the computer.

HOW IT WORKS
From one (black) or three (color) reservoirs inside, ink is thrown onto paper. Each dot of ink is smaller than the diameter of human hair. With amazing precision, the inkjet can fling these dots to render images almost as precisely as traditional (chemical) photographs.

Inkjets use two methods to shoot ink. One is called the thermal bubble or bubble jet technique. Tiny resistors (300 to 600 of them in one print head) heat up when electric current flows through them. As they heat, they heat adjacent ink. The ink vaporizes, expands, and forms a bubble that bursts through a nozzle aimed at the paper.

The piezoelectric method uses crystals (piezoelectric crystals) that vibrate when voltage is applied to them. As they vibrate in one direction, they draw in ink from the reservoir. Vibrating in the other direction, they shoot the ink out a nozzle.

A stepper motor pulls paper from a tray into the printer. A second stepper motor moves the print head precisely back and forth as the paper moves beneath it. Stepper motors don't move like common motors that spin when current is applied. Steppers take one step at a time, typically two to three degrees in size, and thus are ideal for making fine adjustments in position.

INTERESTING FACTS
The cost of inkjet printers has come down so far that you can purchase a printer for less than the cost of the ink it needs.

Laser Printer

BEHAVIOR
It can print your next novel in clarity not possible even a few years ago.

HABITAT
Found adjacent to computers in the home office, these are not typically seen if an inkjet printer is found.

HOW IT WORKS
Laser printers evolved from photocopiers and operate much the same way. The data to be printed comes from the

computer to the printer where it is stored until the page is ready for printing. A laser draws the images of letters or graphics on a rotating drum and this creates tiny negatively charged areas on the drum. Toner, a messy dry plastic powder that takes the place of ink, has a positive charge and adheres to the places with negative charge. The drum rotates to come in contact with a sheet of paper that has an even stronger negative charge. (Notice how paper and especially overhead transparencies have such a strong electrostatic charge when they come out of a laser printer.)

The paper, with toner adhering to it, passes through two rollers that press the toner and heat it. The paper comes out with the toner fixed as a printed page.

Although more expensive than inkjet printers, toner is much less expensive than are inkjet replacement cartridges; so laser printers can be more economical to operate, depending on your useage.

Paper Shredder

BEHAVIOR

It cuts paper documents and unwanted credit cards into strips for safe discarding.

HABITAT

Most shredders have their own dedicated trash bin into which the cut paper falls. They are located near electrical outlets in an out-of-the-way area of the home office.

HOW IT WORKS

An optical sensor detects when paper is thrust into the shredder. The sensor turns on the motor that drives the blades. Some shredders also have manual on/off switches.

Paper passes between two parallel bars each supporting disks with sharp teeth. The disks interweave, leaving little space for the paper to slide through except between the teeth. The motor spins the bars and the toothed disks pull the paper in, chew it up, and spit it out into the bin below.

More expensive shredders cut both horizontally and vertically. This provides an even greater degree of security, but at a price. These shredders cost more and require more maintenance.

You can "unshred" shred paper. It's laborious, but can be done, not unlike an enormous jigsaw puzzle. So if you don't want anyone to be able to read the documents you're shredding, mix the shreds in different waste bins.

Hub or Ethernet Hub

BEHAVIOR
Allows two or more computers in a home to share files and to connect to the Internet.

HABITAT
The hub itself sits atop or adjacent to the computer and near the DSL or cable router.

HOW IT WORKS
Special cables, called CAT 5 cables, connect the router to the hub and connect each of the computers to be networked to the hub. These cables are capable of handling much faster rates of data transmission than are telephone cables (CAT 1 cables). CAT 5 cable has an unshielded twisted pair of copper wires with RJ-45 connectors at each end. These connectors look like, but are larger than, the standard RJ-11 telephone jack.

Hubs permit connections between computers and to the Internet. They receive signals through one of the ports and repeat that signal to all the other computers that are connected. It is called a multiport repeater. Hubs can only handle one stream of data at a time, so if two streams of data arrive at the same time, the sending computers must resolve the conflict through a programmed protocol. Hubs also allow several computers to share peripheral devices, such as printers.

UNIQUE CHARACTERISTICS
Ethernet hubs transmit data at either 10 Mbps or 100 Mbps, the latter being called "fast" Ethernet. Mbps is a unit of transmission speed, a million bits of data per second. Most come with four, five, or eight ports for computers and peripherals to connect to.

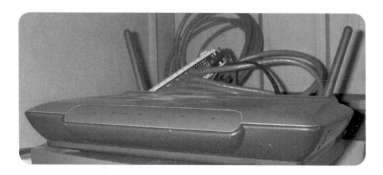

Wireless Access Point Router

BEHAVIOR
Allows computers to exchange information with sites on the Internet without having a wire connection to a phone or cable wire.

HABITAT
Easier than running wires through a home, setting up a wireless router makes every room a potential office. The router is located adjacent to the desktop computer. Laptops or other desktops (with Wi-Fi cards) can be located anywhere in the house or even outside.

HOW IT WORKS
Wireless networking, also called 802.11 networking to differentiate it from other wireless systems, requires an access point that is connected to the Internet by wire. It transmits radio frequency signals to wireless-enabled computers up to 100 meters away. The 802.11 refers to the Institute of Electrical and Electronics Engineers (IEEE) standards specified for this technology, which uses 2.4 GHz and 5 GHz (gigahertz, or billion cycles per second) frequencies to transmit and receive data.

The router performs three separate functions. It is a hub and connects to the DSL or cable modem. It contains a firewall to provide a primary level of security to keep intruders out of your computers. And it provides wireless connections to up to 30 computers.

UTILITY ROOM, BASEMENT, AND GARAGE

UNFINISHED BASEMENTS, GARAGES, and utility rooms are the most fun places in a house to explore. Everything is visible. You can find the machinery that heats your water, washes your clothes, and powers the entire house. Although the furnace is found here, I've included it in the next chapter, along with other components of the air heating and cooling system.

Circuit Panel

BEHAVIOR

It houses the circuit breakers and the electrical connections between the wires bringing electricity into your house from outside and the wires that carry it throughout your house.

HABITAT

It is built into the wall or mounted on the inside wall directly opposite the electric meter that is on an outside wall. Wires carrying the electric power run from the meter outside to the circuit panel inside.

HOW IT WORKS

The drop wire (or underground wire) from the utility pole is really three wires: two wires connect to opposite ends of a transformer (hung on the utility pole or mounted on the ground) and one wire from the center of the transformer (providing the neutral connection). Most appliances in your house need to connect to one of the two wires tapped to the ends of the transformer that supply 120 volts. To complete the circuit, the appliance is also connected to the neutral wire so electricity flows from one "hot" side through the appliance to the neutral side. There is a fourth wire, the ground, that connects the circuit panel itself to the ground through a thick grounding strap made of woven wire.

Inside the panel electricity is parceled out to different circuits. Open the panel door and read the labels. You are safe to open the door, but removing the panel cover (by removing the screws) exposes wires carrying electricity. You might have one circuit for the living room and dining room, but a different circuit or circuits for the kitchen, which uses a lot of electricity. The utility room has a circuit, but the dryer in the utility room also has its own circuit due to its high electric demands. Appliances that use a lot of electricity—air conditioners, refrigerators, dryers, hot water heaters, ovens, etc.—typically have their own circuit breakers.

Dryers and some other appliances may use 240 volts instead of 120 volts. To receive 240 volts, the dryer is connected to the two hot wires coming from opposite ends of the transformer. Each one supplies 120 volts, so combined the total is 240 volts. All appliances that use this higher voltage have different plugs and heavier wires.

UNIQUE CHARACTERISTICS

If a circuit breaker "breaks," that is, if you lose electricity to part of your house, check first to find what tripped the breaker. Someone might have just overloaded the circuit by turning on a hair dryer when several other appliances were drawing power from the same circuit. In this case, plug something into a different circuit in a different part of the house. If this wasn't the case, look for a possible short circuit among the wires and plugs feeding appliances on the blown circuit. When you have removed the problem, find the blown breaker—you can usually see that it is out of position. You can also wiggle it; it will feel less secure than the other breakers. Pull the errant breaker to the "Off" position and then toggle it back to the "On" position. If it trips out again, you know there is a problem in the circuit or appliance—a problem that requires attention.

Circuit Breaker

BEHAVIOR
A circuit breaker protects property and lives by automatically shutting off the flow of electricity if the flow becomes excessive.

HABITAT
It resides inside the electrical distribution panel (see previous entry).

HOW IT WORKS
Older home systems (and electrical systems found in cars and some appliances) use fuses. A fuse is a device that is designed to fail. It carries electricity across a thin metal wire that will melt and break if the current flow is too large. Thus if a short circuit occurs or if too much electricity is being drawn through one circuit, the wire inside the fuse will break, shutting off the circuit.

Short circuits can occur if two electrical wires, one carrying electric energy and the other connected to the neutral or ground, come in contact with each other. For example, if the insulation is damaged on the two wires that power a lamp and the wires touch, the circuit is "shorted" and the fuse "blows." Once blown, a fuse has to be replaced with a new one.

Newer homes have circuit breakers, which are mechanical devices that "trip" when a circuit is overloaded or shorted. Excessive electrical current energizes an electromagnet inside that pulls a lever to shut the circuit. To reset the circuit breaker (after the problem has been fixed) you merely open the circuit case, push the switch to its full "Off" position, and return the switch to its "On" position.

One master circuit breaker protects all the other circuits. It is at the top center of the distribution panel. Each circuit in the house has its own circuit breaker. These are mounted in two columns inside the distribution panel.

Ground Wire

BEHAVIOR
Protects the home and everyone inside from electrical shock.

HABITAT
The ground wire connects a metal rod driven into the ground adjacent to the house foundation with the ground "bus" (connection) inside the electric distribution panel.

HOW IT WORKS
The ground wire provides a high conductivity outlet for electricity so you don't get shocked. When connected to the case of an appliance it ensures that you won't get shocked if you touch the appliance while an errant wire inside, or a wire with frayed or damaged insulation, carrying electricity also touches the appliance. The ground shunts electrical charge to keep your safe.

> "Ground" means an electrical path to earth. In the early days of electric circuits, it was discovered that telegraph operations didn't have to have two wires strung between stations to form a circuit. Instead, one wire could carry the electrical signal, provided that the two stations each connected the other side of the circuit to the ground.

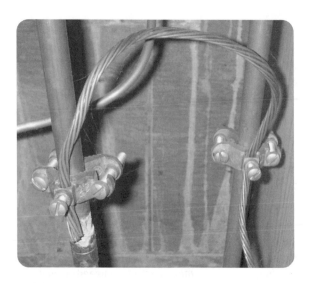

Cold Water Pipe Grounding Wire

BEHAVIOR
Provides electrical grounding for the plumbing to prevent you from being shocked.

HABITAT
Somewhere on the copper pipes that carry cold water, there is a "ground" connection with a bare copper wire. The connection is probably located near the circuit breaker.

HOW IT WORKS
The ground wire protects you from shocks. If the copper pipes in your house somehow came in contact with the electrical circuit, the ground wire would carry the current to ground, preventing you from getting shocked.

Freezer

BEHAVIOR
Keeps food at temperatures well below freezing, typically at 0°F (−18°C).

HABITAT
A freezer, unlike a refrigerator, doesn't often command the nicer real estate of a home. It is often relegated to the garage, utility room, basement, or pantry.

HOW IT WORKS
A freezer works as a refrigerator does. It moves heat from inside the insulated box to outside the box by circulating a fluid through pipes.

When the fluid is passing through pipes in the box, it absorbs heat from the ice cream and frozen pizza. When it is passing through the coils you see on the back of the freezer, it transfers heat to the surroundings.

The process for freezing foods commercially was developed by Clarence Birdseye. He figured out how to fast-freeze food, eliminating the buildup of frost and reducing the damage to the cell walls of food. As foods cool, water inside freezes, forming large ice crystals. These crystals can break open food cells, allowing nutrients and flavor to escape. His method prevented the large crystals from forming. He sold his patents to a company that later became General Foods. It launched a brand of frozen food named in his honor: Birds Eye.

Clothes Washing Machine

BEHAVIOR
Turns out sparkling clean clothes with little human labor.

HABITAT
Washing machines are usually hidden out of sight. They reside in basements, garages, utility rooms, and apartment closets.

HOW IT WORKS
These are marvels of engineering. Basically, the washing machine is a steel tub within a steel tub, surrounded by a steel shell. The inner tub spins to force water out through the holes in its sides. It also has a central agitator that moves clothes and detergent-laden water back and forth to flush out mud, yesterday's jam, and body odors.

An electromechanical timer controls the whole contraption. As the timer moves, it makes and breaks electrical contacts that open valves (letting water in and out) and switches on pumps (to circulate water and to suck water out) and a motor (to spin the drum and turn the agitator). The motor also powers the pumps. There are safety switches to shut off the washer should the lid open in mid-cycle or should the water level rise too high, and a switch to stop the spin cycle if the load is unbalanced, setting off one of the world's most annoying buzzers.

The inner steel tub holds the clothes, and the outer tub keeps the water in. The motor that turns the agitator back and forth is the same motor that spins the inner tub to expel the water. The pump that circulates water is the same pump that empties the water out of the tubs. Very cool engineering.

People have been washing clothes for centuries but using washing machines only for a few years. The first machines appeared in the 18th century and gradually improved. But it wasn't until the 1960s that automatic washing machines became commonplace. One more reason to celebrate the '60s.

Clothes Dryer

BEHAVIOR

Blows warm, dry air through a steel tub while spinning the clothes. It lets you extract half-damp clothes when you're late for a meeting.

HABITAT

Next to the clothes washing machine.

HOW IT WORKS

Warm air holds more moisture than does cold air, so by blowing warm air amongst the freshly washed clothes it picks up moisture. With enough time, usually a few minutes longer than patience allows, the clothes are dry.

The clothes dryer is a steel tub that spins, an electric motor, a heating element, and a control. The motor both spins the tub and turns the fan that draws air into and pushes it out of the dryer. The motor drives the tub with a long belt. Inside the dryer you can see the belt and how it is wrapped around the drive pulley. Also, you can see that a second pulley on a spring provides tension on the belt.

Air is pulled in from the room, past the heating element, and into the tub. There it picks up moisture in the form of water vapor. The air is pulled out through holes in the inside of the front door or the back of the machine, down into the lint trap, past the fan, and out the long, flexible tube (with the giant slinky inside) to the outside.

A simple electromechanical timer controls the operation. As it turns it engages controls that turn on and off the several functions. Soon, one wonders, will this be replaced with a microcontroller that will sense the humidity in the tub and shut itself off when the clothes are ready?

Dryer Outlet

BEHAVIOR
Permits electric dryers to be connected to electrical power.

HABITAT
Found in utility rooms and garages where homebuilders expect the residents will install a clothes dryer.

HOW IT WORKS
The unique configuration of the metal pins and their corresponding openings in the outlet ensure that only the correct appliances can be connected to this type of outlet.

Nearly all of the other outlets in a house provide 120 volts of alternating current. However, electric dryers require 240 volts. Two of the three prongs supply power from opposite ends of the transformer (on the utility pole or ground pad outside). Each end provides 120 volts compared to the center tap, so combining both ends gives 240 volts. The third leg of the outlet/plug is for ground.

Newer houses have four-prong connectors for electric dryers. The top prong is round and provides a separate connection to ground. The bottom prong is L-shaped and connects to the neutral or center tap of the transformer. The prongs on each side connect to the two end taps of the transformer.

Gas-powered dryers use gas to provide the heat and electricity to power only the motor that spins the drum. These dryers plug into a standard 120-volt outlet.

Garage Door Opener

BEHAVIOR
Pulls open the garage door to let you drive in.

HABITAT
The motor and controls are usually mounted on a small platform attached to the ceiling of the garage. Newer houses have electrical outlets mounted into the ceilings so the garage door opener can plug in.

HOW IT WORKS
The motor doesn't lift the weight of the door, which can be quite heavy. Instead it gets help from springs that hold much of the door weight. The springs can attach either to the door itself on each side (extension springs) or to a roller bar atop the door (tension spring), connected by a metal wire to the door. The force of a tension spring is not to be messed with.

Some systems use a belt drive and others a screw or chain drive that connects the motor to the door. The openers are designed to reverse direction when closing if they encounter an object (say, your car or bike) in the way. Garage doors kill or disable about four children a year. To prevent anyone or anything from being crushed, garage doors have either a pressure sensor, which detects resistance to the door's downward movement, or an optical sensor, which detects something in the way a few inches above the floor. It's a good idea to periodically test your garage door to ensure that it is working properly and that it will reverse if it encounters a person or object.

Before cars became popular, there wasn't much need for garages or garage doors. The first automobiles were stored in carriage houses or in public livery stables that could house dozens of cars and provide maintenance. By 1910 car owners were demanding a new structure for their cars. Sears sold prefabricated garages to meet this need.

The first overhead garage door was built by Overhead Door Corporation in 1921. This was a solid door that tipped to open; the top tipped inward and the bottom tipped outward.

Iron

BEHAVIOR
Seldom used today, when pressed into service it removes wrinkles from clothing.

HABITAT
Mostly it sits unused in a closet or atop an ironing board.

HOW IT WORKS
First things first: irons don't have iron in them. They once did, but not today. Today they are made of plastic with an aluminum soleplate (the ironing surface).

A heating element in the base of the iron heats the soleplate. Water from the reservoir drips down into a vapor chamber where it is heated and converted into steam. The steam flows out through holes in the soleplate.

The magic of irons is that they flatten out wrinkled clothes. In washing, the long molecules in fibers curl up. Heating them (and using steam on cotton fabrics) relaxes the bonds between adjacent fibers. The flat surface of the iron presses or aligns the fibers and they hold that alignment when the fabric cools.

Irons date back to at least the 17th century when they were heated by the coals of a burning fire. Later models carried burning coals to provide a steadier source of heat. Henry Seely invented the electric iron in 1882. Inventors have improved irons many times by coating the bottom surface with nonstick Teflon and adding steaming capabilities.

Laundry Chute

BEHAVIOR

Lets gravity do its thing to transport the soiled clothes to the laundry area.

HABITAT

Found in multistory homes. The terminus is in the utility or laundry room. The launch area is an opening in a hallway wall near the bedrooms.

HOW IT WORKS

More often the subject of great stories rather than misguided adventures, the laundry chute holds its own fascination. Being able to communicate and transport instantly between distant parts of the house is too cool to ignore.

Unlike trash chutes in high-rise apartments, the laundry chute holds fewer dangers. Lift the lid or open the door and send your grass-stained socks and sweaty gym clothes away . . . hopefully for someone else to pick up and deal with.

> Add a counterbalanced elevator to the chute and you have a dumbwaiter. Not often found in homes today (more often in libraries, offices, and restaurants), dumbwaiters are a classic piece of home architecture. Squeeze inside one and lift yourself—or have someone else pull you up—to the next floor. Adventures like this provide lifelong memories. Thank you, Grandma Chapin.

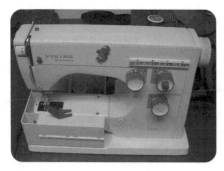

Sewing Machine

BEHAVIOR

It repairs torn and worn clothing and creates new clothes from fabric. It binds two or more pieces of cloth material together with thread.

HABITAT

If in operation it is often found in project rooms, utility rooms, unused bedrooms, or in the dining room.

HOW IT WORKS

What makes the sewing machine effective is the lock stitch. A loop of thread is pushed through the two pieces of fabric. A separate piece of thread stored below the fabric feeds through the loop to prevent the loop from sliding out the hole.

The needle is driven by an electric motor and moves only up and down through an opening in a metal plate. It carries with it a thread fed from a spool that sits atop the machine. Underneath the metal plate is a second spool, or bobbin, of thread. As the needle pokes through the two pieces of cloth, a hook below the plate catches the thread it carries. The hook pulls the thread around the bobbin (holding the second thread) and as the needle is extracted, the loop catches the other thread coming off the bobbin. The bobbin thread keeps the loop from pulling back through the cloth. A "feed dog" (metal plate, with teeth, that moves back and forth) pulls the fabric through the machine with each up and down stroke of the needle.

All this movement comes from one motor with lots of mechanical parts to affect the right motion and the right timing.

Mousetrap

BEHAVIOR

With blinding speed, the trap's bail snaps shut, indiscriminately on fingers or mice.

HABITAT

A mousetrap is found in the dusty recesses of the garage or basement. It is placed where mice are likely to frequent but people aren't.

HOW IT WORKS

Pay attention closely . . . to avoid serious pain to your pinkies. The mousetrap is an example of an unstable mechanical system. A very small force (merely a touch of the trigger) releases a much greater force. The restraining wire flies off, releasing the spring-power bail.

The restraining wire is a lever that holds the significant force of the spring. A small metal catch attached to the bait holder holds the wire in place. All that is required to release the spring-powered bail is a slight jiggle of the bait holder, which releases the wire.

Although cheese is the favored bait in cartoons, I find that peanut butter is more effective.

Make a better model and the world will beat a path to your door? The trap sold today has withstood the best-laid design plans of people for many years. There are over 4,000 U.S. patents for mousetraps. Sir Hiram Stevens Maxim invented the modern mousetrap. He is better known for inventing the machine gun. Maxim also experimented with steam-powered airplanes and got one off the ground before the Wright Brothers flew. (Alas, steam doesn't provide enough power for the weight it requires, and Maxim wasn't able to solve the problem of flight control.) John Mast created the Victor trap in 1899.

Water Pipe

BEHAVIOR
It carries water into and through the house to bathrooms, kitchens, utility rooms, and miscellaneous faucets.

HABITAT
Water pipes run under flooring. If you have an open ceiling basement, you can follow the water distribution from point of entry (where they enter from the water meter) to pipes carrying water to each faucet or other use.

HOW IT WORKS
Water pipes are made of copper, galvanized steel (in homes made before 1970, gray in color), polybutylene (gray plastic, in homes made from the 1970s to 1990), or CPVC (chlorinated polyvinyl chloride, a cream-colored plastic). When pipes were first used in Roman times, they were made of lead. The word plumbing derives from the Latin word for the element lead. Lead, however, leaches into the water supply and can cause serious medical problems, so lead pipes are no longer used.

UNIQUE CHARACTERISTICS
See if you can trace the flow of water in your house. Start with the pipe that brings water in from the water meter. Once inside, it will branch out to feed the hot water heater and all the cold water faucets. Downstream from the hot water heater you will find pairs of pipes, one hot and one cold (or at least unheated).

Water Hammer

BEHAVIOR

This species is never seen, but often heard. You tend to hear it most when the dishwasher or clothes washer are working. It is the "kaboom" in your water pipes.

HABITAT

It hides inside the walls, but comes exploding out as a shock wave, sure to annoy or awaken you from a nice nap.

HOW IT WORKS

Water hammers occur when an open valve shuts quickly. Water in the pipes is flowing quickly through the valve and then the valve closing stops it abruptly. The impact of the water suddenly stopping moves the entire pipe and you hear the noise as a water hammer.

The valves used in appliances are solenoids (electromagnetic motors) that can open and shut very quickly, leading to water hammers. Typically when people turn off water at a faucet, they gradually reduce the flow and don't generate a water hammer.

INTERESTING FACTS

There are ways to reduce or eliminate a water hammer. Plumbers can install a shock absorber in the pipes. This is a side pipe that accommodates the surge of water when the solenoid closes.

Water Tap for Refrigerator

BEHAVIOR
It connects into a cold water pipe to draw water off for an icemaker. The same type of tap can also feed water into a humidifier.

HABITAT
Look in the pipes directly beneath or behind your refrigerator. Or, follow the plastic tube from the back of your refrigerator to a brass fitting mounted onto a water pipe.

HOW IT WORKS
Installing the tap is easy. You straddle it over a cold water pipe, secure it in position with a pair of nuts and bolts, and then turn the handle of the threaded, self-piercing bit. The bit opens a hole in the pipe and lets water flow through the tap into a connected plastic pipe.

Water Softener

BEHAVIOR
It takes calcium and magnesium out of the water to make the water "soft."

HABITAT
Water softeners can be found in basements, crawl spaces, garages, or closets. To locate it, start where the water pipe enters your house and follow the pipes.

HOW IT WORKS
First, why do you want soft water? Hard water—water with lots of dissolved minerals—tends to clog pipes and water heaters, shortening their lives, and it makes it more difficult to wash. The water

doesn't wash away the soap and dirt, and surfaces stay covered with soap scum. According to the U.S. Geological Survey, 85 percent of homes in the United States have hard water.

The softener replaces the dissolved calcium and magnesium ions (charged particles) with sodium ions. The sodium ions don't have the negative qualities that the other ions have. Inside the softener, water flows over beads covered with sodium ions. Ions are exchanged until the beads are depleted of sodium and covered with magnesium and calcium ions. Then the softener recharges the beads by soaking them in a sodium chloride (salt) solution. The excess brine, mixed with the unwanted calcium and magnesium ions, is drained and the softener is ready to go.

UNIQUE CHARACTERISTICS
If you're not sure if your water is hard or soft, check out your shower and bathtub. If you find soap scum accumulating there, you probably have hard water. Water testing laboratories can measure water hardness, and other properties, for you.

Hot Water Heater

BEHAVIOR
It makes showers so wonderful. It heats water and stores it until needed.

HABITAT
Typically the hot water heater is found in basements or garages. In apartments, it may be found in closets.

HOW IT WORKS
The heater is a glass-lined steel cylinder with a heating element. Heaters are either electric, with two heating elements, or gas, with one heating element and an exhaust pipe to get rid of the combustion

gases. Standing next to a heater you can hear the gas burner start up when water is drawn off.

Cold water flows into the tank under line pressure. Inside the tank, the water is heated. A thermostat controls the heating element. Hot water flows out a pipe. You can easily feel the difference in temperatures of the pipes carrying incoming cold and outgoing hot water. But be careful not to burn your hand.

Electric water heaters have one heating element in the upper part of the tank and a second one lower. Only one heater is on at a time. As hot water is drawn out of the top of the tank, cold water enters at the bottom and the lower heater comes on. Should the demand be so large that the water in the upper tank isn't hot enough, the upper heater comes on.

INTERESTING FACTS

Notice on top of the tank there is a pressure relief valve. If internal pressure exceeds safe levels, this valve will blow rather than allowing the tank to rupture.

Many areas now require hot water heaters to be seismically secured to prevent them from falling over in an earthquake. Check to see if your heater is held in place with metal straps.

Some tanks are covered with thick insulation to reduce heat loss, especially if the heater is located in a garage.

About 17 percent of the energy used in an American home is used to heat water.

Sump Pump

BEHAVIOR
Removes water that collects in a sump to keep a house dry.

HABITAT
Found in basements, in a separate room. A sump is the lowest elevation in the house, designed to collect water that seeps in from the walls or floor. Sump pumps are only found in wetter climates or in wetter areas.

HOW IT WORKS
Sump pumps draw water up from the sump and dump it into a storm drain or, in some cases, a sewer line. Municipalities prefer (some require) that sump pumps don't empty into sewer lines to prevent wastewater treatment plants from being inundated after heavy rains.

Sump pumps may be located at the bottom of the sump, immersed in water. These are submersible pumps. Their advantage is that they can directly suck up the water; they don't require an intake pipe. But they are harder to service. Other sump pumps are mounted above the water on a platform or pedestal, making them easier to maintain.

Sump pumps operate on automatic switches that detect water levels. Most use a float; as the float rises with the water level in the sump, the attached arm moves electrical contacts. Submersible pumps have pressure-activated switches. As water rises, the switch exerts increasing pressure on a diaphragm and pressure sensor.

Because sump pump failures are common and costly, some homes also have water level alarms. A sensor in the sump detects high water levels and a radio transmits a signal to a receiver with a built-in alarm.

Sumps need to be covered so debris and pets don't fall in. They also need periodic cleaning to get rid of grit and dirt. Power should always be disconnected before servicing a sump pump.

Cable Connector

It connects the incoming TV cable with all of the cables that provide service throughout the house.

Cables often run under the flooring of the main floor, so they are visible when standing in the basement.

The cable signals come into the house either from a connection box on a utility pole outside or from an underground line. You may be able to find the cable mounted along the outside of your house, inches above ground level.

The cable passes into the house before branching out to provide access in different rooms. Cables connecting computers and television sets meet the wire carrying the incoming signal at a connector. These cable runs are often hung in basements from the floor joists of the higher floor.

Television cables are coaxial cables and are easy to spot. They are round and white. Inside, at the center of the cable, is a conducting wire. It is sheathed in an insulator, which is surrounded by a second conducting wire, which is wrapped in the white insulation you see from the outside. Coaxial cables have their own connectors, very different from other electrical or telephone connectors.

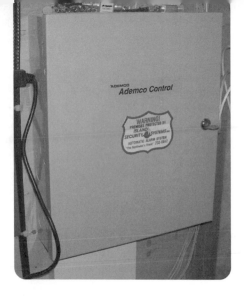

Alarm Control Box

BEHAVIOR
Receives input signals from the various alarm sensors and sends out alarms.

HABITAT
Located on a wall in the utility room or basement. Requires electrical power, so it must be located near an outlet or be hardwired to the electrical system.

HOW IT WORKS
Signals can come to the control box either through wires or by radio transmission from sensors. Wireless connections require an antenna mounted on the control box.

The control box contains a microprocessor that interprets the alarm signals and routes them appropriately. It can cause a local alarm, bells, buzzers, or lights to activate to scare away an intruder and to announce a fire; or it can send the alarm to a remote location such as an alarm company office.

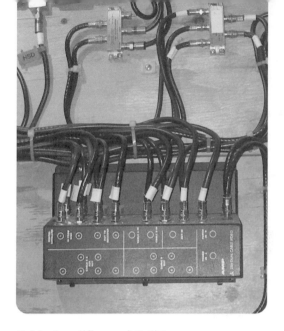

Cable Amplifier and Splitter

BEHAVIOR
Strengthens the cable signal so it can be distributed to many different outlets throughout the house.

HABITAT
Mounted on a wall in the utility room, beneath stairs, or in the garage.

HOW IT WORKS
If you have only two or three rooms with cable access, you don't have one of these devices. However, if you have cable in every room, you might.

The incoming cable signal is only strong enough to provide access to a small number of outlets. To supply more, the signal needs to be amplified. Once strengthened, it can be split amongst the several outlets.

In the photo, the amplifier is the black box. Above the box are some of the splitters that divide the signal among the cable outlets.

Central Vacuum System

BEHAVIOR
This is the machine that creates the vacuum and collects the dust bunnies for the central vacuum system.

HABITAT
Found in garages or utility rooms of homes that have central vacuums.

HOW IT WORKS
The high-speed motor spins impeller blades that draw air from the outlets throughout the house (see chapters 2 and 4). Air and dirt that is sucked in enters either a disposal bag or a vortex collector. When full, the bag or collector is emptied.

HEATING AND AIR CONDITIONING

OF ALL THE ENERGY USED in a home, over half goes toward keeping the house at a comfortable temperature. Maintaining a constant and comfortable temperature indoors is a recent achievement for humans. Central heating and cooling is one of the luxuries that we enjoy but rarely think about. Controlling a large array of machinery spread throughout the house is the thermostat.

Thermostat

BEHAVIOR

Allows you to set a desired temperature and cycles the heater or air conditioner on and off to maintain that temperature.

HABITAT

Thermostats are usually found on a hallway wall, about five feet above the floor. They are located where there is no direct source of heat or cooling, away from outside walls and windows.

HOW IT WORKS

The discovery that metals expand when heated and that they expand at different rates was the genesis for the thermostat. Two strips of different metal are laminated to each other, so that when heated the bimetallic strip bends.

The strip is wound into a coil, spiraling from the center of the thermostat. At the far end of the strip is a glass vial holding a mercury switch. A tiny amount of mercury inside can close the contacts of a switch when the vial is positioned correctly. Electric current flowing through the switch closes a relay that starts the furnace or air conditioner. As the coiled strip moves as a result of temperature change, the position of the glass vial changes, the mercury flows away from the switch, and the furnace or air conditioner stops.

Andrew Ure made the first bimetallic strip thermostat in 1830. But it wasn't until 51 years later that these devices became commonly used, after Charles Hearson improved them. The basic behavior—bending of bimetallic strips as they are heated—was first reported by the famous scientist Sir Humphrey Davy in 1802.

Furnace (Forced Air Heat)

BEHAVIOR
Heats and blows warm air throughout the house.

HABITAT
The furnace is usually found in the basement or utility room. From there it distributes hot air through metal ducts.

HOW IT WORKS
Forced air heating systems use either gas or electricity to heat the air. Gas heating may be the most economical and efficient.

Gas is piped into the furnace either from a gas pipeline or from an outside tank. Under pressure, it flows into the furnace. When the thermostat turns on the relays controlling the furnace, an electric starter ignites the mixture of gas and oxygen (from the air).

After a few seconds the furnace gets hot, and a blower draws air in. Air is pulled in from large vents located in the house where cold air is likely to collect—at floor level in the bottom floor. The blower pulls the air into the furnace and through filters that remove dust. The air is pushed through the furnace in pipes and out, into the ducting made of aluminum.

As air spills out of heat registers in the floor, it readily mixes with room air rather than rising directly to the ceiling. Thus, central heating systems heat a room quickly.

Gas Shut-Off Valve

BEHAVIOR
Allows the gas pipeline to be shut for maintenance or in case of damage to the pipe or furnace.

HABITAT
Gas pipes that feed a furnace and hot water heater have shut-off valves close to the point where the gas pipes enter the appliance.

HOW IT WORKS
The shut-off valve is a hand-operated valve. Gas flows when the handle is aligned with the pipes and doesn't flow when it is perpendicular to the pipes.

This is a ball valve. The handle is attached to a ball inside the metal valve. A hole in the center of the ball allows gas to flow when the handle is aligned with the pipes. Turning the handle rotates the ball so the inner side of the valve covers the hole.

This type of valve is not valuable for regulating the quantity of gas that flows; it is an on/off switch.

Air Filter

BEHAVIOR
Removes dust, pollen, and other particles from the air.

HABITAT
Portable units can be found anywhere, powered by batteries or household electricity. Other units can be integrated into the central air handling system.

HOW IT WORKS
There are several ways to clean or purify air. The simplest is to pass air through a fine filter, like charcoal. The high end of such filters is the HEPA (high efficiency particle arresting) filter. The filter elements in HEPA filters are sheets of microfibers composed of boron silicate. Folded into pleats, these paper-like sheets can capture 99.97 percent of the particles of size 0.3 microns. They capture a higher percentage of particles larger and smaller than 0.3 microns.

A more commonly found method uses electrostatic charge filters. As air moves through the filter, particles collect negative charges from an electrode. The now negatively charged particles are attracted to positively charged plates and they adhere to them. Periodically the plates have to be washed to remove the accumulated particles.

The filter shown here uses screen filters to stop large dust bunnies and electrostatic filters to remove smaller airborne particles.

INTERESTING FACTS
The first electrostatic filter was invented by Dr. Frederick G. Cottrell in 1907. It was and is used to scrub pollutants from industrial smoke.

Air Conditioner

BEHAVIOR
Cools and dries the air circulating through the home.

HABITAT
Found in homes throughout the United States and in many warm countries. At least part of the air conditioning unit is outside. In single homes, it usually sits on a small concrete pad at the side or back of the house. In apartments, small units can be mounted in windows.

HOW IT WORKS
Air conditioners work like refrigerators. A motor compresses a gas, called refrigerant, which increases the gas's temperature. The higher pressure and warmer gas is circulated through pipes outside so it can cool. After cooling, it travels through an expansion valve that reduces its pressure, thus cooling it further. Now the cool refrigerant is circulated in pipes in front of a fan inside the house. Of course, this is the place to stand or sit on a hot day.

Window-mounted units have a single motor that drives the compressor as well as two fans: one to blow outside air over the pipes carrying warm refrigerant and one to blow inside air over the pipes carrying cool refrigerant. Larger units that serve a single home can have an outside component with its own motor and fan, and an inside unit with a motor and fan. Usually this is connected to the ducting used for central heat.

INTERESTING FACTS
William Carrier, an American inventor, installed the first whole-building air conditioning system in a manufacturing plant in Brooklyn in 1903.

Heat Pump

BEHAVIOR

It generates hot or cold air without combustion. It uses electricity to move heat to where it is needed.

HABITAT

Heat pumps are located outside a home, on a small concrete pad. They have pipes leading into the home through holes in the wall.

HOW IT WORKS

Heat pumps work like air conditioners do. They move a fluid through two sets of coils so they provide heat to one set and cooling to the other.

The fluid moves through a condenser where it is compressed, which raises its temperature, and into a set of coils. As it flows through the coils, air is blown through the coils to pick up heat. The fluid then moves through an expansion valve that reduces the pressure and temperature, before passing through another set of coils. Here again, a fan blows air past the coils to be cooled.

In winter, the warmed air is pumped into the house for heating. In summer, the cool air is pumped inside for cooling. A switch controls which temperature air is blown into the house. Heat pumps are an efficient heating source for homes.

INTERESTING FACTS

One of the world's greatest scientists, Lord Kelvin, was the first to describe a heat pump.

Humidifier

It adds water vapor to the air being circulated throughout your house.

Some types connect into the central air handling system and can be found near the furnace. Portable units can be operated in a room and moved to other rooms as needed. Humidifiers are more often found in desert or dry environments where they not only help provide moisture for lungs and nasal passages, but also reduce static electricity build-up that leads to shocks.

There are several ways to get water into the air. In an evaporative humidifier, water is wicked up into a material and a fan blows air through the material. The air blown through the material evaporates the water and carries it away as water vapor. These machines will be cool to the touch since they are evaporating water. Water can come either from a reservoir that you refill (portable units) or from a water tap like the one that supplies water for the icemaker in the refrigerator (for units that are part of the central air system).

Another approach is to boil the water. These are usually small units used for a single room or even bedside for the person who has a cold. They are called vaporizers. The "steam" (actually water droplets) rises toward the ceiling.

An impeller system uses a high-speed motor to turn an impeller blade that throws droplets of water at a screen where they break up. An impeller is a rotating set of arms (similar in appearance to a propeller) that transfers energy from a motor to a fluid to move the fluid. Air moving through the screen picks up the water droplets. You see the stream of water droplets as a cool fog coming out of the machine.

Ultrasonic humidifiers vibrate at frequencies above the range of human hearing. The rapid vibrations knock water into droplets that can drift throughout the room. Since they move at ultrasonic speeds, you don't hear the machine operating.

UNIQUE CHARACTERISTICS
You can differentiate the type of humidifier by what you see and hear. A fast-rising, hot plume of water indicates a vaporizer. A silent system suggests an ultrasonic. Hearing a motor hum tells you that it is either a high-speed impeller (higher frequency sound) or a wicking machine with a lower speed fan (lower frequency sound).

Dehumidifier

BEHAVIOR
Cheaper than an Arizona vacation, they remove moisture from the air.

HABITAT
Found mostly in damp basements and anywhere moist air stagnates.

HOW IT WORKS
It's a lot easier to add water to air than to remove it. Dehumidifiers work like air conditioning systems. A fan draws air into the dehumidifier and past the two sets of coils (one warm and one cool) and back into the room or ducting. A compressor inside the dehumidifier raises the pressure of a refrigerant gas. The pressurization raises the temperature of the gas. The gas flows through pipes that warm the room air before it returns to the room or duct. Passing through a valve that reduces the pressure, the gas expands and cools and passes through another set of pipes. Air coming into the dehumidifier encounters these cool pipes first and vapor condenses on them, because cold air can hold less moisture than warm air. Water droplets form on the chilling coils, then slide down into a collecting tray, where they either drain away or are periodically carried away. Before leaving the dehumidifier, the air is warmed up by passing through the warm coils.

Heating Duct

BEHAVIOR

Carries hot (and cold) air in a central heating system. It also accumulates pennies, dust bunnies, and the miscellaneous small items that kids drop into heat registers.

HABITAT

Ducts are the shiny, metallic, rectangular pipes found along the ceilings of basements, inside walls, and in the flooring of attics. Some are covered with insulation so you might not see the shiny aluminum surface.

HOW IT WORKS

Starting with the furnace, notice the size of the ducts. Quite large at this point, they narrow with each branch to maintain air pressure. The branches lead to heat registers where the air is released into a room.

Ducts also return air from the house to the furnace. Return ducts draw air from cold spots in the lower floors. Vents for return ducts are quite a bit larger than heat registers so air flow is less likely to be blocked.

Duct tape was designed not for use on heating ducts, but as a waterproof sealer for ammunition boxes during World War II. After the war people started using this versatile tape for a variety of tasks, including making repairs on heating ducts. This is when the tape picked up its name. Now most states discourage the use of duct tape on heating ducts, and professional repair people don't use it (but may use a heavier, metallic tape. And just to be confusing they continue to call it duct tape).

Heat Register

BEHAVIOR
Exhausts warm and cold air from furnace and air conditioning in a central, forced air system.

HABITAT
Usually found in the floors where you are most likely to kick them.

HOW IT WORKS
The register is the end of the line for the ducts carrying heated or cooled air in a house. It's the nozzle that sends warm air out into the room on a cold day.

Often registers have covers to direct the airflow. Hot air exiting a register will rise and accumulate along the ceiling, leaving cold air down where the people are. To help diffuse the air, a cover forces it to move horizontally, giving it more time to mix with the cold air at floor level.

Electric Baseboard Heater

BEHAVIOR
Transforms electrical energy into heat by passing electric current through high resistance wire to heat air.

HABITAT
Found along walls at floor level. Found in parts of the country that enjoy lower electrical energy rates.

HOW IT WORKS
Electricity moving through metal wires generates heat from the movement of electrons. Some metals, most notably nickel-chromium compounds, are highly resistive. They are used in toasters and heaters. Nickel-chromium contains about 60 percent nickel, 16 percent chromium, and 24 percent iron.

As the thermostat switches the heater on, current flows through the wires. The wires get warmer and heat the adjacent air.

Baseboard heaters are not as efficient at heating an entire house as is a central heating system. However, if you need to heat only one room, baseboard heaters are a good choice.

UNIQUE CHARACTERISTICS
Baseboard heaters sport fins that transfer the heat from the heating element to the air. Fins provide a large surface area so more air molecules come in contact with the hot surface.

INTERESTING FACTS
Baseboard heaters have a thermal shut-off switch. If the heater becomes dangerously hot (possibly because the flow of air to the heat is blocked), the liquid inside a switch expands and presses against a diaphragm to open the switch. If the switch opens, you may have to reset it by pressing on the reset button where electric wires enter the heater.

Radiator

BEHAVIOR
Heats the house by transferring heat from hot water inside to the surrounding air.

HABITAT
More often found in older homes, radiators are attached to a wall or floor.

HOW IT WORKS
The radiator is the delivery vehicle for heat. Hot water is pumped through the radiator. Fins mounted on the hot water pipe inside the radiator increase the surface area so more air can come in contact with the heat. Convection currents (warm air rising, cold air sinking) carry away warm air and bring in a supply of cooler air to be heated.

Pipes leading to the radiator carry water from 180°F to 240°F that has been heated in a hot water boiler. Valves throughout the system allow for individual radiators to be cut off from the flow of hot water (if a room doesn't need to be heated). Some systems have thermostats in each room that control the valves to adjust the temperature. Return pipes carry the water that has passed through radiators back to the boiler for reheating.

Hot water radiator systems are good at maintaining a steady temperature in a house or larger building, but are slow to respond if you want to change the temperature.

INTERESTING FACTS
Although the Romans circulated warm water through houses for heating, central heating did not appear again until the 18th century in France.

LIGHTING AND ELECTRICAL SYSTEMS

WE LIVE IN THE ELECTRIC AGE. Our appliances and tools require electricity, and we expect it to be available everywhere we are. It's hard to believe that only a century ago electricity wasn't ubiquitously available. A century ago electric motors were just being developed into previously unthought-of products. Electric signals were just being amplified and adjusted to launch the age of radio, television, computers, and all the other marvelous technology we enjoy today.

Electricity is so common that at least in developed countries, few people give it a thought except when paying the monthly bill. The flow of electricity depends on a complex infrastructure that can stretch thousands of miles—from mining raw materials to generating plants to homes and offices, having traveled through long transmission lines, transformers, and a host of other gizmos. For a system as complex as electric power is, it is a wonder that it is so reliable. Except for the occasional explosion of a nearby fuse on a utility pole and the resulting service outage, it is almost always available.

Most power in the United States comes from the burning of fossil fuels: coal, oil, and gas. Nuclear fuels and hydropower make up the bulk of the remaining capacity, with small percentages coming from geothermal, solar, and wind power. However the electricity is generated, it is transmitted along high-voltage lines to local substations where the voltage is greatly reduced and sent out on distribution lines that bring it to our neighborhoods. Still, the voltage is too high to use in homes so transformers (usually) mounted on utility poles drop the voltage to what we need inside: 120V and 240V. (For the story of how electricity gets to the utility pole outside, see *A Field Guide to Roadside Technology*.)

Three wires leave the transformer (atop the pole, or at ground level feeding underground delivery wires) and carry +120 volts, -120 volts, and 0 volt current that alternates at 60 cycles per second. Most other countries deliver electricity at different voltages and at different frequencies.

The final step of the long journey is through a drop wire from a utility pole or through a wire buried underground to the house. Of course before you get to use the electricity, you have to pay for it. Electricity passes through an electric meter mounted on the side of your house. From there, it goes to a distribution panel filled with circuit breakers (see page 134). Each circuit in your home can be shut off manually or automatically (if the current becomes excessively high) at the distribution panel.

Most household circuits carry 120 volts. Either of the two powered lines coming into your house, plus the third or neutral wire, can supply this voltage. For those appliances requiring 240 volts, connections to all three wires are required.

Not everything electrical requires 120 volts or alternating current. Many appliances powered by the electricity coming in from the utility pole need electricity at a different voltage or in a different form. A transformer changes the voltages (usually downward).

The first use of electricity in homes was lighting. There are few things more eerie than walking into a dark house. We find comfort, as well as safety and utility, in lighting our homes, and we have many ways of doing it. Up to the end of the 19th century, light was created on demand only by burning combustible fuels. Edison's improvement of the electrical generation of light switched the world on to convenient and safe lighting. It also led the way to dozens of improvements and new methods to extract light from flowing electrons.

Electric Outlet

BEHAVIOR
This device allows you to connect appliances that use electricity to the electric circuit in your house.

HABITAT
They're everywhere, except of course, where you need them. Even a few years ago builders didn't anticipate today's demand for electricity, so older homes have noticeably fewer outlets. Newer homes have several sets of outlets in each room. Most are located about a foot above the floor, mounted in the wall.

HOW IT WORKS
One slot in the outlet is "hot" and delivers 120 volts, while the other slot is neutral, or at 0 volts. Electricity flows from the hot side, through the appliance, to the neutral side. The third hole in the outlet, the round hole, required in the United States since 1965, is "ground."

Appliance plugs have two flat metal pins that parallel each other and fit into hot and neutral slots of outlets. The third arm on plugs is a round metal pin, slightly longer than the other two. This round connector provides a ground connection, and its length ensures that a circuit or appliance is grounded before it is energized with current. Grounding ensures that should the hot wire somehow come in contact with the case of the appliance, someone touching the case won't get shocked. Instead, the errant electric current will flow to the ground wire.

UNIQUE CHARACTERISTICS
Many appliances come with electrical plugs that have different sized pins. A three-pin plug has a round ground pin, but the other two pins may be different sizes as well. This ensures that you will insert the plug only in the desired way. And that keeps the hot side of the circuit connected to the switch inside the appliance, providing one more measure of safety.

Look at the ends of the flat metal pins of an electric cord. You'll see a small hole in each pin. When the plug is inserted into an outlet, two corresponding dimples inside the outlet fit into these holes to help hold the plug in place.

INTERESTING FACTS
Harvey Hubbell invented the electric outlet and plug in 1904. He also invented the pull-chain switch for electric lights.

Other Electrical Outlets

Other outlets provide voltages different than the standard U.S. 120 volts.

You can find different outlets in the utility room and kitchen. If you travel to other countries you can also find a variety of outlets.

HOW IT WORKS
Power-hungry appliances often use 240 volts instead of 120 volts. The plugs still have three (or sometimes four) connectors, but the pins are larger and are set at angles that prevent you from inserting them in the wrong outlets. Two pins each supply 120 volts to give a total of 240 volts, and the third connects to ground.

Other countries use different voltage electricity in their homes and in some places use a different frequency. Several European countries have developed different forms for plugs and outlets. Some have the pins set at angles instead of parallel to each other. Some outlets have switches on them so you can energize or de-energize everything connected through that outlet.

INTERESTING FACTS
In the United States, alternating current (AC) power is generated at 60 cycles per second. Most of the rest of the world uses power at 50 cycles per second. Before traveling to a new country with your electric appliances, check to see what voltage and frequency they use and if you can purchase an adapter that will fit your appliance.

Switch

BEHAVIOR

A switch allows you to control the flow of electricity to lights and other appliances (ceiling fans, for example).

HABITAT

You find switches at chest height along the inside wall of rooms. You can tell the location as it is often dirty from fingerprints of errant hands trying to find the switch in the dark.

HOW IT WORKS

The switch lever is made of plastic, which does not conduct electricity. On the inside of the switch, the other end of the lever pushes on a metal yoke that makes contact between two sets of terminals. A spring holds the switch lever in place in both the "On" and "Off" positions.

Some switches don't have the snap-action of a spring switch and instead are mercury switches. These noiseless switches rotate a drum of mercury that provides the path for electricity to flow through. In the "Off" position, a non-conducting divider keeps the mercury from contacting both sets of terminals. Rotated to the "On" position, a hole in the divider allows the mercury on each side to connect.

Over time you develop an innate understanding for where switches will be found. Traveling to different countries can leave you groping for the switches, as they are often not where they're "supposed" to be. Sometimes they are located on walls outside the rooms they serve and sometimes at knee level rather than chest level.

Two-Way Switch

BEHAVIOR

Allows you to light or extinguish a light from either end of a room. Think about that for a moment and then try to draw a diagram of the wiring that would allow the switches to work. Pretty tricky.

HABITAT

Two-way switches are found mounted on walls in rooms that have two entrances. You might want to turn on the light as you enter the room from either side, so a switch is available at each entrance.

HOW IT WORKS

One side of the light is connected to power and the other side is connected to one of the switches. The two switches are wired together so that when they both are in the up position or both in the down position they complete the circuit and the light comes on.

The switches themselves are different from ordinary electrical switches: they have an extra set of contacts so they can be connected. The first position (there is no set "On" or "Off" position) in the first switch is wired to the first position in the second switch. And the second position of each switch is connected.

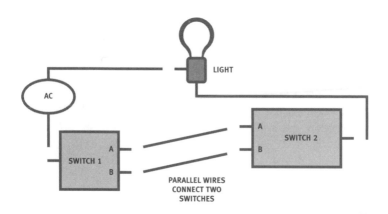

One switch is connected to the light, and the other switch is connected to a wire providing electrical power. When both switches are in the first position, current flows into the first switch, across to the second switch, and out to the light. The same thing happens when both are in the second position, but the current is blocked when one switch is up and the other is down.

Dimmer Switch

BEHAVIOR
Rather than turning a light on or off fully, a dimmer switch allows you to get intermediate lighting levels.

HABITAT
Dimmer switches can be used anywhere, but most often are found in dining rooms (are we having pot roast or a romantic dinner?). They also are used to illuminate paintings or photographs.

HOW IT WORKS
Variable resistors are how they don't work. In physics lab you can get the voltage you want by using a variable resistor, but this approach wastes power that ends up as heat. You don't want your wall heating up, and you don't want to pay for electricity going to waste.

So instead, dimmer switches have a circuit inside that switches on and off 120 times each second. Each time the alternating current changes from positive voltage to negative and from negative back to positive, the dimmer turns off. When it senses zero voltage, it turns off. And it stays off for a length of time dependent on how far you twist the dial.

The knob you turn is a variable resistor. It is connected to a capacitor (an electric device that stores energy). As you turn the knob, you change the resistance and that changes how quickly the capacitor fills up with electric charge. When it fills, it discharges and that triggers an electronic component (a triac) to fire. Thus by turning the knob you change where in the cycle of alternating current the triac closes, allowing current to pass through the switch to light the light.

INTERESTING FACTS
Why don't you see the lights flicker as they turn on and off 120 times a second? Although the power turns on and off, the bulb stays hot and illuminates the room. However, with a dimmer switch, its illumination is at a lower level of light because it is receiving less power.

Cheaper dimmer switches may interfere with radio reception. The rapid switching on and off generates an electromagnetic signal that can interfere. Better switches have an extra capacitor to dampen the electromagnetic fields.

Light Timer

BEHAVIOR
Turns on lights automatically to fool potential burglars into thinking you are home, and to give you a lighted home to return to after dark.

HABITAT
Located throughout the house near electric outlets and lamps. They are inserted between a lamp and the outlet.

HOW IT WORKS
The timer is a clock with switches. The clock has a small electric motor that spins at the frequency of the electric current. The frequency of electric energy, 60 cycles per second (hertz) in the United States, is controlled at the generating plant. (If you visit a generating plant, look for the gauges that show the frequency of the output current to see how close to 60 cycles per second it is). The motor spins and turns a dial.

As the dial turns, it flips switches or relays that control the output power. You can adjust the time of day that the lights come on and off by changing the position of the relay arms.

Incandescent Lightbulb

BEHAVIOR
Passing electric current through these ingenious devices transforms electrical energy into light (and heat).

HABITAT
In flashlights, lamps, and (hopefully) everywhere you need to see in the dark.

HOW IT WORKS
Sir Humphrey Davy discovered in 1801 that passing a strong enough electric current through a piece of metal could heat it sufficiently that it gives off light. Blacksmiths knew that heated metals give off colors, but Davy discovered that electricity could provide the heat. The problem in making a lightbulb was to find a filament material that could be used over a long period of time and to find a way to exclude oxygen from the bulb so the filament would not oxidize or burn. Early bulbs had a vacuum inside; later inert gases (nitrogen, argon, or krypton) filled the bulbs.

INTERESTING FACTS
In the United States we credit Thomas Edison with inventing the lightbulb. Elsewhere in the world credit is often given to both Edison and Englishman Sir Joseph Swan, who independently created a working lightbulb at about the same time as Edison. In actuality, neither invented the lightbulb; they made it work well enough to become a useful product. And Edison developed the electrical system, from generating stations to distribution wires to lamps, that made the lightbulb worthwhile at home. The development of the electric lightbulb is not the "Edison did it" story, but instead is a convoluted story of innovation, science, and legal battles.

Three-Way Bulb

BEHAVIOR
Provides three levels of lighting: low, medium, and high.

HABITAT
Living room lamps are a particular haunt of this bulb, but they could be used anywhere provided that they screw into a three-way lamp fixture.

HOW IT WORKS
You know that a standard lightbulb has one filament that heats up to emit light. How many filaments do you think it takes to make a three-way bulb? Three is the wrong answer. One filament lights for the low light level and a different filament lights for the medium. To get the high light level, both filaments are heated.

UNIQUE CHARACTERISTICS
If you have a burned-out three-way bulb, you can reverse engineer it to understand how it works. Be careful: there is danger of cutting yourself on broken glass! Drape an old T-shirt over the bulb that is resting on a durable, hard surface. Use a small hammer to tap the bulb, just hard enough to break the glass, but not to shatter it to lightbulb heaven. Peel back the T-shirt and peer inside. Look for the filaments. How many filaments are there in a three-way bulb? Can you see a difference in the filaments? Use the T-shirt to gather up the scraps of glass and dispose of it all carefully.

Fluorescent Light

BEHAVIOR
Produces light at lower cost with less heat generated than incandescent bulbs.

HABITAT
You find fluorescents in the home where light might be needed for long periods of time and in places where the bulbs are hidden from view. Often they are mounted behind valances or architectural decoration. Home shops, large closets, and kitchens are examples of places to look.

HOW IT WORKS
What's really cool is that light generated by the gas inside the tube is invisible. If the tube were uncoated and you plugged it in you would still be in the dark! However, the inside of the tube is coated with phosphors (material that absorbs light and then reradiates it). The light generated by the gas is in the ultraviolet range (hence invisible), but it excites the phosphors to give off light in the visible range. In other words, it is the coating on the inside of the glass tube that provides visible light.

When the light is turned on, the starter heats up and emits electrons into the gas tube. The fast-traveling electrons collide with gas atoms and cause their electrons to take on more energy. The higher energy state is unstable, and the electrons quickly give up energy in the form of light.

UNIQUE CHARACTERISTICS

They flicker. Can you see the light flicker? It turns on and off 120 times each second, much too fast to really see, but you might notice a visual blur. You won't see that with an incandescent bulb, although both types use 60 cycles per second alternating current. The incandescent bulb filament can't cool fast enough to stop emitting light. That is, it has thermal inertia that keeps the light coming.

INTERESTING FACTS

The hum you sometimes hear from a fluorescent light is usually the ballast. The ballast is what starts the lighting operation by using line voltage to send electrons flying through the gas. Since the line or house voltage alternates at 60 cycles per second (in the United States), it goes through zero volts twice a second (once rising to 120 volts and once falling to −120 volts). Thus the hum you hear is 120 cycles per second (close to the note B2). Although you might not detect the flicker of the light going on and off 120 times each second, some people report that it gives them headaches.

Compact Fluorescent Lightbulb

BEHAVIOR
Screws into the socket for a incandescent bulb, but operates as a fluorescent. It provides more light for less energy cost.

HABITAT
Found in discriminating light fixtures everywhere. Compact fluorescent bulbs are becoming more popular.

HOW IT WORKS
Compact fluorescent lamps are fluorescent bulbs with a built-in ballast or starter. As in traditional fluorescent lights, the ballast shoots out electrons into the gas-filled tube, adding energy to the gas atoms. As the atoms return to their normal state, they give off the excess energy as light. The light generated is ultraviolet—great for a sunburn, but otherwise invisible. To make visible light, the inside of the glass is coated with phosphors that absorb the ultraviolet light and reradiate light at visible wavelength.

Compact fluorescents should not be used with dimmer switches, electronic timers, or photocell switches. Using such devices can shorten the life of the bulb, and in some cases cause other damage. There are some compact fluorescents made to be used with dimmer switches, so check the package carefully.

INTERESTING FACTS
An incandescent light is only about 10 percent efficient. Ninety percent of the energy consumed goes to generating heat or light outside the visible range. Fluorescent bulbs are about three to four times as efficient and can last 10 times longer. Although compact fluorescents cost more to purchase, they save money in the long run.

Light Emitting Diode (LED)

BEHAVIOR
These are tiny sources of light, used in many appliances to indicate the mode of operation.

HABITAT
They're in your computer, printer, stereo, digital clocks, remote control devices, telephone, battery chargers, and maybe your thermostat. Turn off all the lights in your house on a dark night and the remaining dots of brightness, LEDs (or possibly LCDs, liquid crystal displays), will stand out.

HOW IT WORKS
A light emitting diode is a semiconductor, or electronic component, consisting of a sandwich of materials that conduct electricity under some conditions. A diode lets electrons pass through in one direction, but not the other.

As current passes through an LED, it forces electrons to jump across the gaps between the two different types of material. Only highly energized electrons make the jump and when they land, they give off their excess energy as a photon (or packet) of light. So the process of generating light in an LED is fundamentally different than in incandescent (heating metal wire) or fluorescent lights (shooting electrons through gas).

UNIQUE CHARACTERISTICS

If you have a high-power hand lens or low-power microscope (10X), look closely at an LED. About 90 percent of the bulb is a plastic envelope. In the center is a tiny cup with the actual diode.

If you can see the entire LED, including its leads or legs, you'll notice that they aren't the same length. Because LEDs pass electricity only in one direction, you need to know which is the right direction before installing them in a circuit. The longer leg is the positive (+) side, or the side that connects to the power supply, and the shorter leg connects to the ground.

INTERESTING FACTS

LEDs require little energy to generate light, and they last a long time.

Heat Lamp

BEHAVIOR
Radiates wonderful warmth when it's a cold morning and you're stepping out of a shower.

HABITAT
Most often found inside a bathroom to provide some warmth for your skin. However, heat lamps are also used to keep food hot and used in saunas.

HOW IT WORKS
A heat lamp generates light from electrical power, just as incandescent lights do. However, much of the energy it radiates is in the infrared range, beyond the range that humans can see. Infrared light or radiation warms the skin.

Some houses have infrared heat. Wires in the ceiling radiate heat downward to warm people, pets, and things, while passing through the air. You feel warm, but the air around you can be cool until it warms up by contact (conduction) with objects.

UNIQUE CHARACTERISTICS
Because you don't want to inadvertently leave a heat lamp on and possibly ignite a fire, installed heat lamps operate with a timer switch. You rotate the dial to set the time, and the switch, turned by a spring, shuts the circuit off.

Grow Light

BEHAVIOR
A grow light provides full spectrum lighting needed by plants.

HABITAT
In greenhouses or areas devoted to growing plants. Also found above enclosures for reptile pets or aquariums.

HOW IT WORKS
Full spectrum lights provide light with wavelengths ranging from infrared to ultraviolet, while incandescent bulbs provide energy in the infrared and visible range. A grow light is installed in fluorescent fixtures. The bulb contains mercury vapor or halogen to generate the full spectrum of radiation. Since the ultraviolet light can be dangerous to humans and other animals, it should be not be used for general lighting.

Holiday Lights

BEHAVIOR
They add decorative sparkle to a Christmas tree, front porch, and holiday wreath.

HABITAT
These are seasonal lights (except on my neighbors' house where they stay up all year, although unlighted most of the time).

HOW IT WORKS
The tiny bulbs are 2.5-volt incandescent bulbs. The bulbs are in strands of 50 so the total voltage drop is 2.5 volts times 50, which is close enough to the voltage sup-

plied by your wall outlet, 120 volts. If the strand has 100 or 150 bulbs in it, you'll notice that there are three wires instead of two supplying power. These longer strands are multiples of 50-strand light circuits, where each circuit is connected to household electricity.

UNIQUE CHARACTERISTICS
In much older holiday light strings, if one bulb failed, the whole string would fail. This happened because electricity had to pass through each bulb in the series; one broken bulb interrupted the circuit and stopped the flow of electricity. Newer bulbs have a shunt—a piece of metal—that connects the two legs of the filament. The shunt allows power to flow past the bulb even if it has burned out.

Blinking lights use a device similar to a thermostat. Inside the bulb is a bimetallic strip (two different metals attached to each other). As current flows through the strip it heats, and because the two metals expand at different rates, it bends. Bending makes it lose contact with the filament, thus extinguishing the light. As it cools, it returns to its initial shape, making contact and lighting the light.

Black Light

BEHAVIOR
It's dark, but it lights.

HABITAT
Found in totally groovy places during parties.

HOW IT WORKS
Black lights are fluorescent lightbulbs that don't emit visible light. Most of the light generated is in the ultraviolet range. Thus you don't see light when you look at a black light.

What you do see is the effects of that light radiating throughout a room. As the ultraviolet light hits objects, it energizes them. Some immediately re-emit light at a longer, visible wavelength. This is called fluorescence. Some materials will absorb the energy and slowly give off light. This is phosphorescence.

The desired effect at party time is to give an eerie appearance as bits of clothing shine brightly in an otherwise dark room.

INTERESTING FACTS
You can conduct some interesting experiments with black lights. Detergents used for washing clothes contain a variety of chemicals to reflect light, so clothes appear whiter or cleaner without actually being whiter or cleaner. Clothes washed in detergents may exhibit more light than those washed in soap. Also, some minerals in rocks exhibit phosphorescence and appear very different in black light than in visible light.

Light Sensor for Outside Lights

BEHAVIOR
It turns on your outside lighting when it gets dark.

HABITAT
Sensors may be mounted on the side of your house, on light poles, or from the eaves of your house. You can identify the sensor by its "eye." It has a plastic or glass cover (about 1/4-inch in diameter) protecting the sensor that is inside the box.

HOW IT WORKS
The sensor is a photo resistor or solar cell. Made of cadmium sulfide, it strongly resists the flow of electricity when exposed to low light levels and weakly resists when light is abundant.

As the sun fades at the end of the day, the photo resistor responds to lower light levels. Its resistance increases to the point that its circuitry switches a relay that turns on the light. Increasing light levels in the morning lower the resistance of the cell, which switches the relay to turn off the light.

Photo resistors are a type of semiconductor. Incident photons (particles of light) provide energy for electrons in the semiconductor to move and conduct electricity.

INTERESTING FACTS
Most devices that change with light levels use this technology—for example, street lights. Also, light detectors used in cameras use photo resistors.

Motion-Activated Light

BEHAVIOR
Turns on flood or other lights when it detects motion.

HABITAT
These lights are often mounted above garage doors, front doors, and driveways to provide lighting when needed without keeping the light on full time.

HOW IT WORKS
There are several ways to detect motion, but the most common is passive infrared. Active infrared sends out a beam of infrared energy to bounce, radar-like, off whatever is in its path. Passive infrared doesn't send out a beam, but receives infrared radiation from everything in its view.

People, animals, and machines give off infrared radiation. In fact, everything that isn't absolutely cold gives off radiation. Motion detectors look for rapid changes in the amount of infrared radiation they see. Someone walking into the field of view will change the level of radiation and cause the detector to turn on the lights.

The material inside the sensor is a crystalline material called pyroelectric material. It gives up electrons when hit by infrared radiation. The flow of electrons inside the device is measured to determine when it has seen motion.

Low-Voltage Light

Provides focused light for ambience.

Most often found in kitchens, but it could be used anywhere.

Low-voltage lighting operates at 12 volts, 24 volts, or 30 volts. A transformer, usually located in the attic or basement, transforms the household current (120V AC) to the voltage required for the lighting system. Low-voltage systems thus require additional wiring to bring the power from the transformer to the lights.

The lights themselves are incandescent bulbs, similar to those found in slide projectors. Typically they provide lower wattage light than a typical incandescent bulb does, but can provide sufficient lighting for many uses. The lower wattage means lower electric use, and lower cost.

The light fixtures usually occur in strings or sets. Each bulb illuminates a small area. The overall effect is attractive lighting.

Emergency Power Transfer Switch

BEHAVIOR
Allows you to switch off the electric grid and receive power from an emergency generator.

HABITAT
Found in homes that experience frequent or long-term power outages. Homes in hurricane-prone areas tend to have these.

HOW IT WORKS
Before turning on an emergency electrical generator, this switch is activated to disconnect the house from power coming in from the electric company. This switch allows the user to power up individual circuits in the house using the emergency generator.

When the tree falls against the electric line and knocks out the power, you need not suffer in darkness. (Of course, candles are a less expensive option.) Throw the switch on your emergency power transfer box and go outside to start up the diesel (or gasoline) engine that powers your generator. Now, while your neighbors huddle together in a dark and cold house, you can crank up the waffle iron or catch the forecast on the Weather Channel.

Generators work the same as do the generators that electric companies employ, though the home version is a tad smaller. A gasoline or diesel engine spins the generator to make electricity.

Electrical Generator

BEHAVIOR
Provides electrical power when the normal utility service is interrupted.

HABITAT
Found in homes where power interruptions are frequent and lengthy.

HOW IT WORKS
Generators can be small gasoline-powered units like the one shown, or can be much larger diesel engine generators. In either case, the internal combustion engine converts the chemical energy of its petroleum fuel into mechanical motion. The circular motion is used to turn an electrical generator, similar to the much larger ones used by electric utilities to generate electricity at power plants.

Most emergency generators require the owner to start them, although automatic self-starters are available. Once the generator is started, it can power some, but probably not all of the electrical needs of the house.

INTERESTING FACTS
These generators are becoming more common in hurricane-prone areas. With a good supply of fuel, the owners can board up their homes and sit out all but the most dangerous storms.

Electrical Heating Tape

BEHAVIOR
Heats water pipes to keep them from freezing.

HABITAT
Found wrapped around exposed water pipes in climates that experience hard freezes.

HOW IT WORKS
Plugged into an electrical outlet, the heating tape warms up. It transfers its heat to the pipe it surrounds. The tape contains high resistance wires that convert electrical current into heat.

INTERESTING FACTS
Electrical heating wires and tape are also used on roofs to melt snow and ice.

Transformer

BEHAVIOR

Changes the voltage from house current (120V AC) to the voltage required for a specific use.

HABITAT

In the basement, mounted under the flooring of the first floor, or in the attic.

HOW IT WORKS

Some systems inside the home require alternating current at voltages other than the standard 120 volts. The front doorbell and thermostat are two systems that require different voltages. Transformers are devices that change the voltage of alternating current.

Transformers have two sets of wires wound around an iron core or ring. One set of wires has more wraps than the other, and the ratio between the number of wraps determines the ratio between the voltages of incoming and outgoing electric power. To supply a device with 120V AC, the transformer would have a 10:1 ratio of wrappings, with the higher number on the 120V side.

APARTMENT BUILDING

10

LIVING IN A HIGH-RISE or even a low-rise multifamily building offers some interesting technology that single-family homes don't have. Most obvious is how to move people up and down and how to keep them safe.

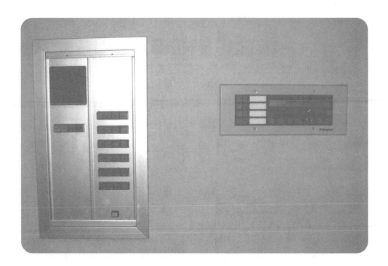

Emergency Enunciator

BEHAVIOR
This announces emergencies and helps locate where the emergency is occurring.

HABITAT
Emergency enunciators are located in the entryway of apartment buildings and office buildings, and at monitoring stations in nursing homes.

HOW IT WORKS
When sensors and alarms in a building are activated, their location is indicated on the enunciator. Emergency responders can check the enunciator to find where the problem is.

UNIQUE CHARACTERISTICS
In any large building, look near the lobby, elevators, or front doors for a panel that lists locations throughout the building.

Fire Alarm

BEHAVIOR
Detects fires and warns residents.

HABITAT
Detectors are often mounted in the ceilings along hallways. Human-activated alarms are mounted high on walls.

HOW IT WORKS
Alarms can be tripped manually by pulling a lever or by one of several types of smoke sensors. Only building personnel can reset manual levers once they've been pulled. Staying in the "pulled" position allows fire teams to know which pull station was activated.

Sensors can detect smoke, heat, or the flow of water through an activated sprinkler head. If a fire sprinkler detects heat and turns on, the flow of water through the pipes can alert the alarm system that a fire is present.

Fire Sprinkler

BEHAVIOR
Automatically releases water when heat rises above a threshold level.

HABITAT
Found in multifamily buildings and occasionally in single-family homes. Located in the ceilings and mounted high on walls.

HOW IT WORKS
Each sprinkler has its own heat detector. When temperatures rise above the threshold, a metal link melts and the sprinkler releases water in the pipes. Other types of detectors use a chemical reaction or the breaking of a small glass capsule to trigger the flow of water. A pump or series of pumps respond to the drop in water pressure in the piping and turn on to maintain the water pressure.

Elevator

BEHAVIOR
Gives quick vertical transportation to the floors in an apartment building.

HABITAT
Found in multistory buildings.

HOW IT WORKS
Two types of elevators are used: one for buildings with only a few floors and the other for high-rise apartments. Smaller buildings typically employ hydraulic elevators. The elevator cab rides atop a piston housed in a shaft that lies underground. To rise, pressure is exerted on hydraulic fluid beneath the piston, which forces it to rise. You can often

hear the pump start to operate when the cab moves. These systems are similar to the hydraulic lifts used in service stations.

In taller buildings, pulley systems are preferred over hydraulic systems. Pulleys and motors are mounted in a small shack on the roof. The weight of the cab is counterbalanced with a set of metal or concrete weights so the motor doesn't have to lift the entire weight of the cab and riders.

Regardless of type, all elevators have a call system and a computer program for figuring out where the cab should to go to respond most efficiently to calls.

Although known as the inventor of the elevator, Elisha Otis didn't invent the elevator; he invented the safety brake for elevators and then used it to create the safety elevator. Before his invention few buildings rose above five floors, because people didn't want to walk higher and elevators without safety brakes weren't safe. His invention changed the landscape of cities throughout the world, horizontally and vertically.

Trash Chute

BEHAVIOR
Allows residents to deposit trash without having to to carry it to street-level cans or bins.

HABITAT
Found in high-rise apartments.

HOW IT WORKS
Enclosed chutes are built into a building so residents can drop trash (and not get hit by trash falling from higher floors). Trash is collected typically in the basement and removed. Secure doors keep any smells from wafting up from the basement collection area into the living areas and suppress fires that could start in materials lodged in the chute.

> Chutes are typically too small to allow human entry, but in December 2003 a Chicago man got wedged in a trash chute. He told rescuing fire-fighters that he had been trying to retrieve his gold watch when he got stuck. The firefighters had to knock down part of the wall to extract him.

Security Camera

BEHAVIOR

Allows security guards to view areas throughout an apartment building so they can alert police or the fire department if needed.

HABITAT

Cameras are usually installed at corners of buildings, near elevators, in the lobby, and in other areas of egress. The monitoring station where the cameras are viewed is at the guard station of the building or at an offsite security company location.

HOW IT WORKS

Most security cameras are fixed in position, but some may be moved for better viewing by the observing guards. Unlike broadcast television, the signals are carried by wires from the camera to the monitor and any recording equipment. These are a type of closed-circuit television.

Cameras are usually mounted high enough to prevent being tampered with. Light enters through the lens and is detected by a charged coupled device (CCD). The CCD has an array of tiny capacitors (electronic components that hold charges). Light hitting each capacitor causes it to hold a charge in proportion to the amount of light that hits it. Thus the image projected by the lens is transformed into an array of electric charges. This array is read by an electronics circuit that sends it to a monitor for viewing.

Monitoring is done real-time, but the video is also taped and kept until it is determined not to be needed (no incidents occurred).

Magnetic Card Reader

BEHAVIOR
Keeps out nonresidents. Residents swipe their magnetic cards and, upon verification, the exclusion gate arm lifts or the gate opens so they can enter the parking lot or housing development.

HABITAT
Apartments and some private homes have security gates controlled by card readers.

HOW IT WORKS
The plastic card has a strip of magnetic material, like magnetic tape from a cassette tape recorder. The tape is plastic film embedded with tiny iron oxide particles that act like magnets. Electromagnetic encoders align the tiny magnets on the card so a read head can read them. If exposed to strong magnetic fields, the particles can be remagnetized into a new pattern, thus losing the code . . . and you don't get in!

In addition to magnetic card readers, security systems can use optically encoded cards (bar codes) or radio frequency identification (RFID) technology. The RFID sends out a radio signal identifying the owner so the receiver will open the gate.

PATIO, PORCH, ROOF, AND OUTSIDE

OUTSIDE YOUR HOME, on the ground, on the roof, and on the side of your house are many devices that work for you. Many deliver or measure essential resources that keep your house operating. Others cook the steaks of the summer barbeque season and keep the lawn green. Take a look outside to see all the strange gizmos.

Electric Meter

BEHAVIOR
It measures the amount of electricity you use in quantities of kilowatt-hours. A kilowatt-hour is the power consumed by a device that uses 1,000 watts that is run for one hour.

HABITAT
On the outside wall where power lines come either up from underground or down from a utility pole. All the electric power inside a house has to pass through the electric meter before coming in.

HOW IT WORKS
Older meters are electrical-mechanical, meaning they have a small motor inside to turn the dials. The more electricity you use, the faster the motor spins. Newer meters are electronic and don't have mechanical or moving parts.

Traditionally a "meter reader" had to go to each house and business each month to read the meter. Now utilities are adding radio transmitters to the meters so usage is radioed to their accounting office, largely eliminating the need for someone to come to your house to read the meters. If you don't see or hear a meter reader, check out utility or light poles in your neighborhood for a small box with one small antenna sticking up and two small antennas sticking down. This device radios your electric usage to an accounting office of the electric company.

UNIQUE CHARACTERISTICS

Try reading your meter. Start with the number on the left side. Record the number that the left dial has passed. To the right of this number, write down the number that the second dial has passed. And, to the right of that number, record the number that the third dial has passed. Repeat this procedure for the remaining dials. Note that the dials turn in opposite directions, leading to confusion about how to read them.

This number, by itself, doesn't give you enough information to tell how much electricity you have been using, because the meter didn't start from zero at the start of the month. Keep the number you just recorded and check the meter a few days or weeks later, recording the new number. Subtracting the first reading from the more recent reading will give you the kilowatt-hours of power you have used in that time interval. However, some meters may require you to multiply the number you calculated by a factor. Look for the factor written on the faceplate of the meter. If you don't see one, you don't need to do this extra step.

Multiplying the number of kilowatt-hours you use in a month by the cost per kilowatt-hour will give you an estimate of your electric bill. Of course, to get the same cost as your electric company computes, you will have to take readings the same days it does. And, increasingly, electric companies charge different rates throughout the day, so multiplying the kilowatt-hours by an average cost per kilowatt-hour will only give you an approximate cost.

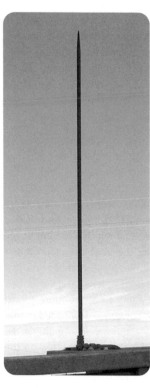

Lightning Rod

BEHAVIOR
Carries dangerous electrical charges from lightning bolts from the roof to the ground.

HABITAT
Placed along the ridgeline or along the highest point of a roof. The rods appear more frequently on larger homes and on homes in states that experience a lot of electric storms.

HOW IT WORKS
A lightning rod does not attract lightning, or at least that's not its purpose. It gives lightning an easy path to ground, rather than passing through your house and possibly starting a fire or causing other damage.

The high-voltage charges that build up in an electric storm find their way to ground through a bolt of lightning. If that bolt hits a structure in its path to ground, it can damage or destroy the structure. By providing an electrical path around the structure, lightning rods and the wires that connect them to ground protect the structure.

Benjamin Franklin, prolific inventor and all-around creative guy, came up with the idea for lightning rods and installed the first one on his Philadelphia home in 1749.

Barbecue Grill

BEHAVIOR
Cooks steaks, ribs, hot dogs, and hamburgers on open flame or glowing coals.

HABITAT
Found on the porches and patios of homes throughout the United States. Many hibernate during cooler months or are at least kept under tarps or wraps.

HOW IT WORKS
The center of many a summer party, the barbecue grill burns wood, charcoal, or gas to provide heat and open flames. Gas

(liquid propane or natural gas) is most convenient. Liquid propane comes in pressurized cylinders that can be refilled at service stations, and natural gas is delivered to homes through pipelines. A small pipe carries the gas to the barbecue grill.

Opening the valve releases natural gas through a vent. The gas spreads out through a series of burning ports where it mixes with air and ignites. The operator can start the flames either with a fire starter or match, or with a grill starter mounted on the grill.

The grill starter uses piezoelectric material. Depressing the button compresses a spring. As it passes a point, the spring is released and hits the piezoelectric material. The physical impact on the piezoelectric material generates an electric voltage potential that manifests itself as a spark.

Grills with covers can be used to bake food. Damp wood chips are sometimes added to the fire to provide smoke for the food to absorb. Chicken, pork, and beef are popular entrees for grilling.

Solar Heater

BEHAVIOR
It heats water by absorbing solar radiation.

HABITAT
Most often found on the south-facing roof of a house. You can identify the flat panels resting above the roof. In some cases, the pipes that carry the water are enclosed in a cylindrical case that focuses the solar rays on the pipes.

HOW IT WORKS
Solar water heaters are either open or closed systems. In a closed system, a fluid (with a low freezing point) is pumped through the solar collector on the roof and into a hot water tank. In the tank, the fluid heats the water that will be piped throughout the house.

In an open system, the water you use is pumped through the solar collector and back inside the house. Because this system circulates water rather than an antifreeze fluid, this type of system has to be drained before freezing temperatures occur to prevent pipe rupture.

Water Meter

BEHAVIOR

Measures how much water is used in a house. In almost all cities residents pay for water based on how much they use. Cities measure water consumption with the meters and charge customers based on the reading.

HABITAT

Your water meter resides somewhere between the water main (probably running under the street) and the pipes that bring water into your house. In many neighborhoods, the water meter is buried in a small vault located along property lines. Look for a metallic cover that swings open.

HOW IT WORKS

Water moving through the meter spins an impeller (a rotor designed to turn with the water flow). The impeller is connected to a series of gears that drive the counter you see on the face of the meter. As the flow of water increases, the impeller spins faster.

UNIQUE CHARACTERISTICS

Increasingly water utilities are reading water meters electronically rather than manually. You can figure out which system your utility uses: if you never see a meter reader walking through your neighborhood, that is the first clue. The second clue is that you can find a small transmitter attached to your water meter.

With the radio transmission system, the meter reader drives through a neighborhood. A computer data retrieval system inside the car or truck sends a message to each water meter asking it to send its measurement. The computer records all the data. Back at the water department, the computer downloads the data and sends it to the billing department, which sends out the monthly bills. What used to take several days of work can be done in a few hours with the radio system.

Clean Out

BEHAVIOR
Gives access to a sewer line so it can be cleared.

HABITAT
The clean out is found where the household's septic lines connect to the sewer line.

HOW IT WORKS
With a large wrench, a plumber unscrews the lid. Then he inserts the end of a metal "snake" into the pipe and forces it towards the main sewer line in an attempt to unclog the line.

Roto-Rooter uses a rotating cutting head to trim tree roots and unclog the pipes. A motor spins the cutting head. A flexible drive cable inside a metal jacket supplies the turning power. Water flushed down the pipe carries away the debris.

> The Roto-Rooter device was invented in the late 1920s by Samuel Blanc. He cobbled the first machine together from parts from a washing machine and a pair of roller skates.

Stack Vent

BEHAVIOR
It vents gases and admits air into the plumbing system, allowing it to remain at atmospheric pressure (preventing vacuums from forming).

HABITAT
Vent pipes stick up above the roof. A house may have one or more vents. If bathrooms are located in one part of the house (which they are to save on the cost of running more water and waste water pipes), the stack vent will probably be located above them. Another stack will be located above the kitchen plumbing.

HOW IT WORKS
Stack vents sit atop a long pipe that ties in to all the drains in the house. As wastewater flows down drains it pushes gases out of the way; the gases escape up through the stack vent. As wastewater drains into the sewer line, its volume has to be replaced to prevent a potentially blocking vacuum from occurring. The stack vent allows air to enter the drain system to prevent vacuums.

UNIQUE CHARACTERISTICS
Stack vents are approximately four inches in diameter to prevent them from being blocked by leaves or snow. They stick up far enough above the roof to prevent being blocked.

INTERESTING FACTS
You can make a good guess about where bathrooms are by looking for the stack vents. However, architects tend to put these on the back side (away from the street entrance) of a house so they are less visible.

Lawn Sprinkler and Timer

BEHAVIOR
It distributes irrigation water to keep the ornamental foliage healthy.

HABITAT
Homes with owners who are either serious about how their lawn looks or about reducing the labor of hand-watering their lawn.

HOW IT WORKS
Lawn sprinklers can be manually operated or run on a timer. Timers are usually electromechanical systems using an alternating current motor that activates switches. The switches power solenoids, electromagnetic valves that let water through or stop it.

Water pipes are installed underneath grass and along gardens, with vertical risers coming towards the surface. Sprinkler heads are mounted on the riser pipes to direct sprays of water in the desired direction. Some sprinkler heads allow the flow of water to be adjusted so you don't water the concrete sidewalk. Some sprinkler heads are fixed a few inches above ground and others pop up, activated by the pressure of water.

Some sprinklers oscillate or rotate to deliver water to larger areas. These systems have mechanical parts that are powered by the flow of water. Oscillators have a small water wheel inside that spins from the impact of water. The water wheel turns a series of gears that slow the rate of spin to about one revolution per minute. The gears drive a cam that moves the sprinkler head back and forth.

House Protector

It protects the telephone wiring inside your house and you from electrical shocks that could occur from nearby lightning strikes or other electric surges.

HABITAT
Found on an exterior wall, probably near the electric meter.

HOW IT WORKS
The older-style protector has a carbon fuse; the newer style uses a gas tube fuse. A surge of electricity burns the carbon fuse, allowing a grounding wire to come in contact with the "tip" and "ring," the two wires that carry phone conversations. The surge in electricity is shunted to the ground, thus protecting anyone using the phone. When this occurs, phone service is interrupted until the fuse is replaced. The newer gas tube protector diverts surges to ground, but then returns to normal operation without the need for a service call.

Even older still are ceramic fuses in a 6-inch-long (15 cm) aluminum canister. If these fuses burn out, a repair technician must replace them to restore service.

UNIQUE CHARACTERISTICS
Find where the telephone drop wires come into your house. Look for a small (probably) gray box mounted on the outside of your home that has wires running from the box to the utility pole or underground. It may bear a telephone logo or company name.

Chimney

Supports one or more flues that vent combustion gases to the atmosphere.

HABITAT
Found at or near the ridgeline of the roof of a house.

HOW IT WORKS
Inside the chimney are flues for each combustion source in the house. The fireplace uses one flue, a gas hot water heater uses one, and a gas furnace needs one.

Chimneys get rid of the various noxious gases resulting from burning oil, gas, or wood in the fireplace, furnace, and hot water heater. For a fireplace, the warm gases rising up the flues inside the chimney draw air into the combustion chamber to feed the fire.

The flues inside the chimney may have a liner made of tile or metal. These keep the gases from escaping through the masonry and damaging it.

Chimneys can catch on fire if not cleaned periodically. The need for cleaning is higher when burning high resin wood in a fireplace. Call the chimney sweep!

Attic or Roof Vent

BEHAVIOR
Lets hot air escape from the attic through openings in the roof.

HABITAT
Located near the ridgeline on a roof.

HOW IT WORKS
Warm air rises in the attic towards the high point. Openings in the roof let the air escape to the outside. Vent covers prevent rain, snow, birds, and bats from entering through the openings.

UNIQUE CHARACTERISTICS
If you're on the roof, check out the vents and how they are attached to the roof. Their base is attached so it prevents water flowing down the surface of the roof from running into the opening.

INTERESTING FACTS
Some homes have attic fans that draw air from the house and force it into the attic. From there, the roof vents or louvered openings at the end of the house (on the gable or triangular wall between two parts of a roof) vent the air to the outside. This is a less expensive system for cooling a house than air conditioning.

Gas Meter

It measures and records how much gas you use so you can be billed accordingly.

In some places the gas meter stands out like a pink flamingo in the front yard. A pipe comes up from the ground, into a pressure-reducing valve, then into the big meter, and back into the ground where it connects to the gas system inside your home. In other places, the gas meter is mounted along an outside wall of your house. Walk around the perimeter of your house to find it.

The meter measures how much gas is used and displays the amount in cubic feet. Inside the box are two chambers, each of which is subdivided into two parts. The meter measures how many times gas fills and then empties each of the chambers.

Look at the face of the meter to see if you can figure out how to read it.

Satellite TV Antenna

It collects television broadcasts from one of several satellites and sends the programs to your TV set.

You can see the antennae mounted on roofs and sides of houses. Older, larger antennae are often set on poles in the yard. All face toward the south (in the United States).

To receive reliable TV signals, the transmitter needs to send the signal directly to the receiver. One way to do this is to park a satellite in one location and beam television signals through it.

The way to park a satellite in one place is to launch it into geo-synchronous orbit at 22,300 miles above the equator. At that altitude, satellites revolve around the world in 24 hours, thus keeping pace with the earth spinning below it.

Signals are broadcast to the satellites, amplified onboard, and then broadcast back to earth, a round trip of more than 40,000 miles. The parabolic antenna on the house roof collects the signals from the satellite and sends them into the receiver inside the house.

Signals are compressed to take up less bandwidth and encrypted so if you don't pay for the service, you don't get to use the signal.

Even though the signal may travel 50,000 miles before reaching you, the signals move so fast (at or near the speed of light) that you cannot detect any delay. Sitting at home you will see the big play on the television screen at the same time your cousin in the ballpark sees it, and you will probably hear the sounds before your cousin does.

Television Antenna

BEHAVIOR
Collects television broadcast signals and sends them to your TV set.

HABITAT
Usually mounted on the roof, often supported by the masonry chimney.

HOW IT WORKS
Look closely at a TV antenna. It's not just a random bunch of metal rods sitting on a roof. The standard antenna, called a Yagi-Uda antenna, has three types of arms. One is "folded" into a loop and is called the driven element. A single metal bar at one end of the antenna is called the "reflector," and it is slightly longer than the driver. One or more bars at the opposite end are the "directors." They are slightly shorter than the driven element. The more directors there are, the better the signal pick-up (gain) will be. The director end of the antenna should face the TV station broadcast antenna. The three elements work together to eliminate signals that interfere and to provide the strongest signal possible.

UNIQUE CHARACTERISTICS
If you get close enough to the antenna to see the wire that connects it to the television set, note the type of wire used to carry the signal. It is unlike any other wire in your house. Unlike common electrical wire where the two leads are immediately adjacent to each other, antenna wire has the two leads separated by about a half inch.

Also look for an electric motor attached to the base of the antenna. If present, it rotates the antenna so it can best receive the signals from different television broadcast stations in different geographic areas.

Pool Filter

BEHAVIOR

It removes dirt from the pool water so the pool sparkles in the summer sun.

HABITAT

Found adjacent to swimming pools, which are more often found in warm climates. It can be located outside or concealed inside a shed.

HOW IT WORKS

The filter has a water pump that draws water in from both the pool surface (through skimmer drains) and the bottom drain. Big debris is kept out of the water pipes by screens. The water moves into the filter from the top, down through the filter, and out the bottom. The majority of filters use sand as the filter medium. As the water flows through the sand, particles get lodged in the sand while the water passes through and returns to the pool through another pipe. The accumulating particles of dirt eventually clog the sand medium, slowing down the passage of water and requiring more pressure from the pump to get water through the filter.

Periodically the filter requires backwashing to get rid of the dirt it has collected. Rising pressure, shown on a gauge, alerts the homeowner to the need to backwash. With the pump turned off, the homeowner turns a handle that reverses the flow of water through the filter. Now, water comes up from the bottom of the filter, lifting the dirt and carrying it out the top to the sewer line. A few minutes of backwashing cleans the filter medium, and the filter can be turned on to its normal operation.

Although sand filters are most common, other media are used, too. Some pool filters have cartridges holding thin sheets of material that water can flow through, but dirt particles cannot. Instead of backwashing these cartridges, you remove them from the filter and clean the dirt off with a hose. Diatomaceous earth (or DE, a sedimentary rock formed from the deposits of diatom fossils) filters are used to provide a higher degree of cleaning. Some pool filters use DE filters, as do many aquarium filters.

Although humans have probably floundered about in water since the dawn of our species, organized swimming started around 2,500 B.C. in Egypt. The first heated swimming pool was built in ancient Rome. The industry didn't take off, however, until the 19th century. In the 20th century, as more people started living in warmer climates in the United States, swimming pools became a common home amenity. Flying into Phoenix or another Southwest city, the window-seated flyer is struck by the number of pools (each evaporating huge amounts of water) that dot the landscape.

Pool Heater

Warms the water, making the pool experience even more enjoyable.

Found adjacent to swimming pool filters, not just in cooler climates, but anywhere you find a pool.

Heat can be provided by either the sun or natural gas. You can see solar heaters on the south-facing roofs of houses that have pools. The simplest form involves running many rows of black plastic (ABS) pipe on the roof through which the pool water is pumped. More efficient heaters protect the pipes in an enclosure that, in some cases, helps to reflect and focus the sun's energy onto the pipes.

More common are gas heaters. Pool water travels through copper pipes that are heated by burning natural gas. The photo shows a gas heater.

In most climates pools lose tremendous amounts of heat to the atmosphere. Most of the heat transfer occurs by evaporation, so covering the pool can greatly reduce the cost of heating it.

Pool Cover

BEHAVIOR
It "blankets" the pool to keep in water vapor and heat, and to reduce the loss of pool chemicals.

HABITAT
In use, covers float on the surface of the pool. When not in use, they are either rolled up at one end of the pool, or pushed along rails on each side of the pool to store at one end of the pool.

HOW IT WORKS
Blankets are sheets of plastic that float on the water; some have small air pockets, similar to plastic packing material (the stuff that is so much fun to pop). They allow much of the sun's radiation to pass through to warm the water beneath and block most of the evaporation. Since most of the heat loss occurs through evaporation, a blanket allows a pool to warm up or stay warmer.

Blankets reduce the water loss and loss of chemicals by as much as 70 percent. Thus the savings in reduced heating costs and reduced chemical costs often make blankets a wise investment.

Pool Chlorinator

BEHAVIOR
It automatically adds chlorine and other chemicals to pool water.

HABITAT
Usually it is found adjacent to the pool filter. It feeds chemicals into the water just before it enters the pool filter.

HOW IT WORKS
Chlorine in one of two forms (or bromide) oxidizes organic matter in the water. When introduced into the pool, the chlorine mixes with water to create an acid that burns bacteria and algae to kill them. After swimming in an over-chlorinated pool, your eyes feel the same burn.

The chlorinator has a pump that pushes out a steady stream of chemicals into the water. The advantage of using a chlorinator is that it gradually introduces the chemicals as water cycles through the filter. This tends to dilute the chemicals so they are less noticeable to swimmers.

Timer

BEHAVIOR
Turns on and off appliances such as pool filter pumps.

HABITAT
Found mounted on the sides of pool sheds or houses, adjacent to swimming pools.

HOW IT WORKS
This is an electromechanical timer that uses household current at 60 cycles per second to control the clock. The electricity powers a small alternating current motor. As the motor rotates, it pushes against levers that turn the filter pump on and off.

Remote Thermometer

BEHAVIOR
Senses the ambient temperature, displays it, and sends it by radio waves to an indoor display.

HABITAT
Found on outside walls, hopefully shielded from direct sunlight.

HOW IT WORKS
This battery-powered sensor calculates the temperature based on the resistance of an electronic component called a thermistor. There are several types of thermistors. Simplest to understand is one made of metal oxides. As temperatures rise, more electrons are free to transmit electrical charge and the resistance of the device drops. The circuitry in the sensor measures the resistance of the thermistor and calculates the temperature based on an empirical relationship. A radio transmitter sends the temperature to the indoor display unit.

HOUSEBOAT

SOME PEOPLE LIVE IN HOUSES that don't sit on concrete foundations or on any foundation at all. Instead their homes float on water. Looking out the living room window at a moving river or watching boats and wildlife pass by is a great way to live. And living on the water presents some unique technology.

Floats

BEHAVIOR

Floats support the house above water level and provide the base or foundation for the house.

HABITAT

Found peeking just above water level, largely submerged. Most houseboats are built on huge logs, but to gain additional flotation, closed-cell foam is pushed underneath.

HOW IT WORKS

Logs like these can last over 100 years, provided that they float partially protruding above the water's surface. If left completely submerged, they become waterlogged and can sink. If they float too high, they can rot.

Houseboat owners add flotation when their homes sit too low in the water. Adding a grand piano or a complete collection of *National Geographic* magazines to your floating living room can lower the house. Also, over time, the logs can lose some buoyancy. Adding additional flotation may require hiring divers to position buoyant material underneath the house.

Imagine how hard it is to push a block of flotation beneath the water surface to get it under a house. Divers attach weights to the floats to pull them down and then maneuver them under the house. When in position, they release the weights. The floats rise and the weights sink to the bottom. Crew on board the house or pier use a winch to pull up the weights to attach to the next float.

UNIQUE CHARACTERISTICS

Look at how high a houseboat sits above the water. Will waves wash onto their front porch? Watch the house as a boat passes by to see how much the house rocks.

Piling

BEHAVIOR
It anchors the houseboat in place, while letting it rise and fall with tides and water level.

HABITAT
It is stuck in the mud in the river or lake bottom.

HOW IT WORKS
These steel pilings were pounded into the ground by a pile driver. The houses and docks are connected to pilings so they can move up and down as the water level changes, but can't move away from the moorage.

UNIQUE CHARACTERISTICS
Look at how tall the pilings are. Their height represents the highest water levels that engineers expect to encounter at that location. At flood stage the houseboats will be tugging at the very top of the piling. Imagine your worst day at home, riding downstream in a flood.

Ramp

BEHAVIOR
It allows people to move from shore to dock, regardless of the height of the water level.

HABITAT
Found reaching from the shore parking lot toward the common dock at houseboat marinas.

HOW IT WORKS
The shore end of the ramp is fixed in place, typically by one or two large pins that form a hinge and allow the ramp to pivot as water levels change. The other end of the ramp rests on wheels on the common dock. As water levels rise and the dock floats higher, the lower end of the ramp rolls outward across the dock. The dock must be wide enough to accommodate the ramp at extremely high and low water levels.

UNIQUE CHARACTERISTICS
Sometimes a metal plate covers the dock to spread the weight of the ramp and to make it easier for the wheels to move.

Shock Absorber

BEHAVIOR
Dampens motion of a houseboat while holding it securely.

HABITAT
Found amidst the cables or chains used to anchor the houseboat to the dock or piling.

HOW IT WORKS
Waves and wind move houseboats around, and as a houseboat tugs on the chain or cables holding it in place, it is jerked to a stop. This can provide an uncomfortable experience inside, so springs are added to the mooring hardware to reduce the jerking.

Finger Cuff and Connections

BEHAVIOR
Holds tightly onto electrical cables so that they aren't damaged or disconnected.

HABITAT
Found where electrical cables cross from the pier into the houseboat.

HOW IT WORKS
The metal cuff holds onto the rubber sheathing that protects electrical cables inside. The cuff is connected to the pier to hold it in place. As a houseboat moves, any strain on the electrical cables is borne by the rubber sheathing and not the cables inside.

UNIQUE CHARACTERISTICS
The other utilities, gas, water, and sewage must all have flexible connections since the house and pier move toward and away from each other. Look along the dock to see where they connect.

Sewage Lift

BEHAVIOR
Pumps sewage up and into a municipal sewer line.

HABITAT
Each houseboat has its own small lift, and the moorage has a large lift station, as shown here.

HOW IT WORKS
Sewage flows from high to low to reach the treatment plant. Unlike water, which flows under pressure, sewage flows by gravity. Since the sewage on houseboats starts at the lowest possible elevation—water level or sea level—it has to be pumped up to get it into the sewer lines.

Each houseboat has its own sump and pump, often called its "honey pot." The pump raises effluent to a common sewer line that takes it to a lift or pump. This pumps it up the bank of the river or lake, high enough so it can flow downward into the sewer line.

ABOUT THE AUTHOR

Ed Sobey is science educator, author, and former museum director. He has directed five museums including the national museum of inventing: the National Inventors Hall of Fame. While directing A.C. Gilbert's Discovery Village, Ed founded the National Toy Hall of Fame, now at the Strong Museum in New York.

Ed holds an adjunct faculty position at California State University, Fresno, where he developed Kids Invent Toys, a technology-learning program used by universities and museums in North America and Asia.

INDEX

A Field Guide to
Roadside Technology
BY ED SOBEY

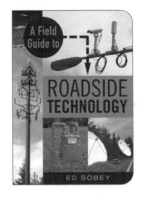

A Selection of the
Scientific American Book Club

"Fun, informative, and easy to use." —SCHOOL LIBRARY JOURNAL

"Proves there really is a travel book out there for everyone."
                                        —DENVER POST

If you've surveyed the modern landscape, you've no doubt wondered what all those towers, utility poles, antennas, and other strange, unnatural devices actually do. *A Field Guide to Roadside Technology* is written just for you. More than 150 devices are grouped according to their "habitats"—along highways and roads, near airports, on utility towers, and more—and each includes a clear photo to make recognition easy. Once the "species" is identified, the entry will tell you its "behavior"—what it does—and how it works, in detail. You'll also learn the history and little-known facts behind the devices you might otherwise take for granted.

$14.95 (CAN $20.95)
ISBN-13: 978-1-55652-609-1
ISBN-10: 1-55652-609-1

Backyard Ballistics

Build Potato Cannons, Paper Match
Rockets, Cincinnati Fire Kites, Tennis Ball
Mortars, and More Dynamite Devices
BY WILLIAM GURSTELLE

A Selection of Quality Paperback Book Club

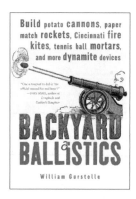

Build potato cannons, paper
match rockets, Cincinnati fire
kites, tennis ball mortars,
and more dynamite devices

BACKYARD
BALLiSTICS

William Gurstelle

"Your inner boy will get a bang out of these 13 devices to build and
shoot in your own back yard, some of them noisy enough to legally
perk up a 4th of July." —THE DALLAS MORNING NEWS

"Would-be rocketeers, take note: Engineer William Gurstelle has
written a book for you." —CHICAGO TRIBUNE

"William Gurstelle . . . is the Felix Grucci of potato projectiles!"
—TIME OUT NEW YORK

This step-by-step guide uses inexpensive household or hardware store materials
to construct awesome ballistic devices. Features clear instructions, diagrams,
and photographs that show how to build projects ranging from the simple—
a match-powered rocket—to the more complex—a scale-model, table-top
catapult—to the offbeat—a tennis ball cannon. With a strong emphasis on
safety, the book also gives tips on troubleshooting, explains the physics behind
the projects, and profiles scientists and extraordinary experimenters.

$16.95 (CAN $25.95)
ISBN-13: 978-1-55652-375-5
ISBN-10: 1-55652-375-0

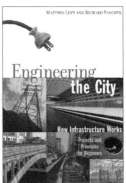

Engineering the City

How Infrastructure Works
Projects and Principles for Beginners
BY MATTHYS LEVY AND RICHARD PANCHYK

"Future engineers, math enthusiasts, and
students seeking ideas for science projects
will all be fascinated by this book."—BOOKLIST

How does a city obtain water, gas, and electricity? Where do these services
come from? How are they transported? The answer is infrastructure, or the
inner, and sometimes invisible, workings of the city. *Engineering the City* tells
the fascinating story of infrastructure as it developed through history along with
the growth of cities. Experiments, games, and construction diagrams show how
these structures are built, how they work, and how they affect the environment
of the city and the land outside it.

$14.95 (CAN $22.95)
ISBN-13: 978-1-55652-419-6
ISBN-10: 1-55652-419-6

CHICAGO REVIEW PRESS

www.chicagoreviewpress.com

Distributed by
Independent Publishers Group
www.ipgbook.com

Available at your local bookstore
or by calling (800) 888-4741